**Lectures in Mathematics
ETH Zürich**
Department of Mathematics
Research Institute of Mathematics

Managing Editor:
Helmut Hofer

Jürgen Jost

Nonpositive Curvature: Geometric and Analytic Aspects

Springer Basel AG

Author's address:

Max-Planck-Institute for
Mathematics in the Sciences
Inselstr. 22–26
D-04103 Leipzig

A CIP catalogue record for this book is available from the Library of Congress,
Washington D.C., USA

Deutsche Bibliothek Cataloging-in-Publication Data
Jost, Jürgen:
Nonpositive curvature : geometric and analytic aspects / Jürgen Jost. -
Basel ; Boston ; Berlin : Birkhäuser, 1997
    (Lectures in mathematics : ETH Zürich)
    ISBN 978-3-7643-5736-8   ISBN 978-3-0348-8918-6 (eBook)
    DOI 10.1007/978-3-0348-8918-6

© 1997 Springer Basel AG
Originally published by Birkhäuser Verlag in 1997
Printed on acid-free paper produced from chlorine-free pulp. TCF ∞

ISBN 978-3-7643-5736-8

9 8 7 6 5 4 3 2 1

Contents

Preface vii

1 Introduction
  1.1   Examples of Riemannian manifolds of negative or
        nonpositive sectional curvature   1
        Appendix to § 1.1: Symmetric spaces of noncompact type 11
  1.2   Mordell and Shafarevitch type problems 19
  1.3   Geometric superrigidity 23

2 Spaces of nonpositive curvature
  2.1   Local properties of Riemannian manifolds of
        nonpositive sectional curvature 33
  2.2   Nonpositive curvature in the sense of Busemann 42
  2.3   Nonpositive curvature in the sense of Alexandrov 54

3 Convex functions and centers of mass
  3.1   Minimizers of convex functions 61
  3.2   Centers of mass 64
  3.3   Convex hulls 67

4 Generalized harmonic maps
  4.1   The definition of generalized harmonic maps 69
  4.2   Minimizers of generalized energy functionals 76

**5 Bochner-Matsushima type identities for harmonic maps
   and rigidity theorems**
  5.1   The Bochner formula for harmonic one-forms
        and harmonic maps 85
  5.2   A Matsushima type formula for harmonic maps 90
  5.3   Geometric superrigidity 95

Bibliography 99
Index 105

Preface

The present book contains the lecture notes from a "Nachdiplomvorlesung", a topics course adressed to Ph.D. students, at the ETH Zürich during the winter term 95/96. Consequently, these notes are arranged according to the requirements of organizing the material for oral exposition, and the level of difficulty and the exposition were adjusted to the audience in Zürich.

The aim of the course was to introduce some geometric and analytic concepts that have been found useful in advancing our understanding of spaces of nonpositive curvature. In particular in recent years, it has been realized that often it is useful for a systematic understanding not to restrict the attention to Riemannian manifolds only, but to consider more general classes of metric spaces of generalized nonpositive curvature. The basic idea is to isolate a property that on one hand can be formulated solely in terms of the distance function and on the other hand is characteristic of nonpositive sectional curvature on a Riemannian manifold, and then to take this property as an axiom for defining a metric space of nonpositive curvature. Such constructions have been put forward by Wald, Alexandrov, Busemann, and others, and they will be systematically explored in Chapter 2. Our focus and treatment will often be different from the existing literature.

In the first Chapter, we consider several classes of examples of Riemannian manifolds of nonpositive curvature, and we explain how conditions about nonpositivity or negativity of curvature can be exploited in various geometric contexts. In particular, we shall explain the rigidity theorem of Mostow and Margulis in a geometric manner. This rigidity theorem will also lead us to the second main topic of these notes, namely the theory of generalized harmonic maps into spaces of nonpositive curvature. This theory has been found very useful in rigidity theory as we shall explain in detail in Chapter 5. We shall systematically develop the ideas of the author for a general framework for harmonic maps, and a general existence result will be obtained in Chapter 4. This will depend on convexity properties and center of mass constructions in spaces of nonpositive curvature developed in Chapter 3.

In parts of this book, we shall employ some basic concepts from Riemannian geometry, like the exponential map, the Levi-Civita connection, the curvature tensor, or Jacobi fields. All this background material can be found in the author's textbook "Riemannian geometry and geometric analysis", Springer, 1995 (2nd edition in preparation).

In these notes, we shall by no means give a complete survey of the theory of spaces or manifolds of nonpositive curvature. The lecture notes [Ba] by Ballmann, for example, contain material that complements the present text.

Acknowledgements

To Igor Nikolaev, I owe some valuable information on the literature about metric spaces of nonpositive curvature. I thank Helmut Hofer for arranging this Nachdiplomvorlesung and for inviting me to contribute the present book. I was stimulated by many remarks and comments from the audience in Zürich. Discussions with Viktor Schroeder were particulary useful. I am grateful to Michel Andenmatten for his help with the redaction of the present text, and to Harald Wenk for his competent typing.

I thank Marc Burger, Martin Deraux, Patrick Ghanaat, Luc Lemaire, Pierre Pansu for useful comments on my manuscript.

Chapter 1

Introduction

1.1 Examples of Riemannian manifolds of negative or nonpositive sectional curvature

In Riemannian geometry, negative curvature usually means negative sectional curvature. Let N be an n-dimensional Riemannian manifold[1]. All Riemannian manifolds will be assumed to be connected and complete unless the contrary is explicitly stated. The scalar product on T_xN, for $x \in N$, defined by the Riemannian metric will be denoted by $\langle \cdot, \cdot \rangle$, the Levi-Civita connection by ∇, and its curvature tensor by $R(\cdot, \cdot)$. The sectional curvature of the plane in T_xN spanned by the linearly independent tangent vectors $X, Y \in T_xN$ is

$$K(X \wedge Y) := \langle R(X,Y)Y, X \rangle \frac{1}{|X \wedge Y|^2}$$

(with $|X \wedge Y|^2 = \langle X, X \rangle \langle Y, Y \rangle - \langle X, Y \rangle^2$).
(Since one always has to worry about sign conventions in Riemannian geometry, let us recall that

$$R(X,Y)Z = \nabla_X \nabla_Y Z - \nabla_Y \nabla_X Z - \nabla_{[X,Y]} Z$$

for $X, Y, Z \in T_xN$.)
N is then said to have negative (nonpositive) sectional curvature if for all $x \in N$ and all linearly independent $X, Y \in T_xN$

$$K(X \wedge Y) < 0 \quad (\leq 0).$$

This is an important concept because on one hand there exist many important classes of examples of Riemannian manifolds that enjoy such a curvature inequality

---

[1]In this §, we shall employ some basic notions from Riemannian geometry. These notions are explained in somewhat more detail in § 2.1 infra. A general reference is [Jo6].

while on the other hand it is strong enough to imply tight topological restrictions and rigidity phenomena.

By way of contrast, if one considers the weaker notion of Ricci curvature, in formulae for $X \in T_x N$

$$\text{Ric}(X, X) = \sum_{i=1}^{n-1} \langle R(X, E_i)E_i, X \rangle$$

where E_i, $i = 1, \ldots n-1$, is an orthonormal basis of the orthogonal complement of X in $T_x N$, then by Lohkamp's theorem, any differentiable manifold of dimension $\geq$ 3 carries a Riemannian metric of negative Ricci curvature. Consequently, negativity of the Ricci curvature does not imply any topological restrictions.

In Hermitian, and in particular in Kähler geometry, however, it is also useful to consider negative or nonpositive holomorphic sectional curvature

$$H(X, X) = \langle R(X, JX)JX, X \rangle$$

where $J : T_x N \to T_x N$ with $J^2 = -\text{id}$ defines the complex structure. A different, though closely related notion in complex geometry is that of hyperbolicity in the sense of Kobayashi. The idea is the following: Let D be the unit disk in $\mathbb{C}$ equipped with its hyperbolic metric d_H. Thus, $D = \{z \in \mathbb{C} : |z| < 1\}$, and d_H is induced by the hyperbolic line element $ds^2 = \frac{dz d\bar{z}}{(1-|z|^2)^2}$. For a complex manifold N and p, $q \in N$, one considers

$$\delta_H^\circ(p, q) := \inf\{d_H(z_0, z_1) : h : D \to N \text{ holomorphic with } h(z_0) = p, \ h(z_1) = q\}.$$

Although δ_H° captures the idea, for technical reasons one needs to replace δ_H° by

$$\delta_H(p, q) := \inf\{\sum_{i=0}^{k} d_H(z_i, z_{i+1}) : h_i : D \to N \text{ holomorphic with}$$
$$h(z_i) = p_i, \ h(z_{i+1}) = p_{i+1}, \ p_0 = p, \ p_k = q, \ i = 0, \ldots, k, \ k \in \mathbb{N}\}.$$

δ_H is a pseudodistance: it satisfies the triangle inequality, but it need not be positive. For example, if $N = \mathbb{C}$, then $\delta_H(p, q) = 0$ for all p, $q \in \mathbb{C}$.

By way of contrast, if N is the hyperbolic unit disk, i.e. D with the metric d_H, it follows from the Schwarz-Pick lemma that

$$\delta_H^\circ = \delta_H = d_H.$$

Definition 1.1.1: A complex manifold N is called *hyperbolic* if

$$\delta_H(p, q) > 0 \qquad \text{whenever} \quad p \neq q \in N.$$

The relation between hyperbolicity and pointwise curvature is given by the Ahlfors-Yau-Royden-Schwarz lemma:

Lemma 1.1.1 ([Ro]): *Let M, N be Hermitian manifolds, M complete with sectional curvature bounded below. Suppose the holomorphic sectional curvature of M is*

bounded below by $\kappa_M \leq 0$, the holomorphic sectional curvature of N bounded
above by $\kappa_N < 0$. Then any nonconstant holomorphic map $h : M \rightarrow N$ satisfies

$$\|dh(z)\|^2 \leq \frac{\kappa_M}{\kappa_N} \qquad \text{for all } z \in M.$$

In particular, if N is a Hermitian manifold with negative holomorphic sectional
curvature (bounded away from 0), w.l.o.g. $\kappa_N = -1$, then any holomorphic map
$h : D \rightarrow N$ is distance nonincreasing. Consequently, δ_M is positive in this case.
Thus, Hermitian manifolds with a negative upper bound for their holomorphic
sectional curvature are hyperbolic. Also, the converse, namely that hyperbolic
manifolds admit Hermitian metrics with negative holomorphic sectional curvature
is probably not very far from being true.
In Riemannian geometry, there are still some other meaningful notions of negative
curvature. The curvature tensor defines a symmetric operator

$$R : \bigwedge^2 T_x N \rightarrow \bigwedge^2 T_x N$$

by putting

$$\langle R(X \wedge Y), V \wedge W \rangle := \langle R(X, Y)W, V \rangle.$$

By definition, R is a negative operator if all its eigenvalues are negative. This,
however, is a very strong condition.

The Riemannian metric $\langle \cdot, \cdot \rangle$ on $T_x N$ can be extended to $T_x N \otimes \mathbb{C}$ as a bilinear
form $(\cdot, \cdot)$ and as a Hermitian metric $\langle \cdot, \cdot \rangle_{\mathbb{C}}$, with the relation

$$\langle Z, W \rangle_{\mathbb{C}} = (Z, \overline{W}).$$

We may define the complexified sectional curvature of $X \wedge Y \in \bigwedge^2 T_x N \otimes \mathbb{C}$ as

$$\frac{1}{|X \wedge Y|_{\mathbb{C}}^2} \langle R(X \wedge Y), X \wedge Y \rangle =: K_{\mathbb{C}}(X \wedge Y).$$

Thus, N has negative sectional curvature if $K_{\mathbb{C}}(X \wedge Y) < 0$ for all real tangent
planes $X \wedge Y$.
$Z \in T_x N \otimes \mathbb{C}$ is called isotropic if $(Z, Z) = 0$. A linear subspace T of $T_x N \otimes \mathbb{C}$ is
called totally isotropic if $(Z, Z) = 0$ for all $Z \in T$.
We say that N has negative isotropic sectional curvature if $K_{\mathbb{C}}(T) < 0$ for all
totally isotropic tangent 2-planes T of N.

Nonpositive sectional curvature implies many local and global properties as we
shall see below. Let us state some of the most important ones.

Theorem 1.1.1: *Let N be a complete Riemannian manifold of nonpositive sectional
curvature. Let p, $q \in N$. Then each homotopy class of curves from p to q contains
precisely one geodesic arc (which then minimizes length in its class). If the sectional*

curvature is even negative, and if N is compact, then each free homotopy class of closed curves contains precisely one closed geodesic.

The existence of geodesics of course holds for any compact Riemannian manifold. Uniqueness, however, in general fails without assuming nonpositive (negative) sectional curvature. Uniqueness also implies that the geodesics of Theorem 1.1.1 depend smoothly on p and q. Uniqueness of geodesics often gives a canonical choice for topological constructions. For example, if $f_0, f_1 : X \to N$ are homotopic maps from some topological space X into the compact Riemannian manifold N, and if $F : X \times [0,1] \to N$ is a homotopy with $F(x,0) = f_0(x)$, $F(x,1) = f_1(x)$ for all $x \in X$, we let

$$\gamma_x : [0,1] \to N$$

be the geodesic arc from $f_0(x)$ to $f_1(x)$ in the homotopy class defined by $F(x,t)$, $t \in [0,1]$, and parametrized proportionally to arclength. The canonical homotopy between f_0 and f_1 is then obtained by putting

$$f_t(x) := \gamma_x(t).$$

f_t, $t \in [0,1]$, is called the geodesic homotopy between f_0 and f_1.

The following result is usually called the Hadamard-Cartan-theorem.

Theorem 1.1.2: *Any complete, simply connected, n-dimensional Riemannian manifold Y of nonpositive sectional curvature is diffeomorphic to $\mathbb{R}^n$.*

Such a manifold N is often called a Hadamard manifold.
Since for any manifold N, any continuous map $\varphi : S^m \to N$ ($m \geq 2$; S^m denotes the m-dimensional sphere) can be lifted to the universal cover, $\widetilde{\varphi} : S^m \to \widetilde{N}$, as S^m is simply connected for $m \geq 2$, we obtain

Corollary 1.1.1: *For any Riemannian manifold N of nonpositive sectional curvature, the higher homotopy groups $\pi_m(N)$ ($m \geq 2$) are trivial.*

By way of contrast, other topological groups associated with a Riemannian manifold of nonpositive sectional curvature can greatly vary as we shall see from a consideration of some examples.

1) The n-dimensional torus T^n equipped with its Euclidean metric has vanishing curvature.

   Its fundamental group $\pi_1(T^n)$ is $\mathbb{Z}^n$. Actually, for any compact Riemanninan manifold N of nonpositive sectional curvature, the rank of any Abelian subgroup of $\pi_1(N)$ is at most n, and if the curvature is even negative, its rank is precisly 1 (if the Abelian subgroup is nontrivial) by Preissmann's theorem. Since the fundamental group of a manifold of nonpositive curvature cannot contain elements of finite order, in the negative curvature case therefore all nontrivial Abelian subgroups of $\pi_1(N)$ are isomorphic to $\mathbb{Z}$.

T^n admits selfmaps $\varphi : T^n \to T^n$ of arbitrary degree. Also, the mapping class group of T^n, the group of homotopy classes of diffeomorphisms $\psi :$ $T^n \to T^n$, is infinite for $n \geq 2$. The automorphism group of T^n, i.e. the group of isometries $i : T^n \to T^n$, is a nontrivial smooth (nondiscrete) group, with identity component given by T^n itself.

Finally, T^n admits moduli, i.e. T^n can be continuously deformed in the class of Euclidean tori, or, in other words, there exist nontrivial continuous families of Euclidean tori. On the other hand, one still observes some kind of rigidity; namely, any Riemannian metric of nonpositive sectional curvature on a topological torus is automatically flat, i.e. has vanishing curvature.

2) Any compact Riemann surface Σ of genus $g \geq 2$ carries a Riemannian metric of sectional curvature $\equiv -1$.

$\pi_1(\Sigma)$ is a nonabelian group with $2g$ generators. The first homology group is $H^1(\Sigma, \mathbb{Z}) = \mathbb{Z}^{2g}$. The first Betti number b_1 of a surface of negative curvature can thus be arbitrarily large.

Any selfmap $\varphi : \Sigma \to \Sigma$ has

$$|\deg \varphi| \leq 1.$$

The mapping class group Γ_g of Σ is again infinite.

The automorphism group $I(\Sigma)$ of Σ, however, is finite, with the explicit estimate of Hurwitz:

$$\sharp I(\Sigma) \leq 84(g - 1).$$

This estimate is sharp: For infinitely many g, there exists such a Σ of genus g with $\sharp I(\Sigma) = 84(g - 1)$.

Again, Σ admits moduli. The corresponding "Riemann moduli space" M_g can be identified as T_g / Γ_g, where the "Teichmüller space" T_g is diffeomorphic to $\mathbb{R}^{6g-6}$. Also, the constant curvature metric of Σ can be deformed inside the class of nonpositive curvature metrics to one of nonconstant negative curvature.

3) Compact locally symmetric spaces N of dimension ≥ 3 and of noncompact type[2] and rank 1, i.e. compact quotients of real, complex, quaternionic hyperbolic space ($H_{\mathbb{R}}$, $H_{\mathbb{C}}$, $H_{\mathbb{Q}}$, resp.)(of dimension ≥ 3) or the hyperbolic Cayley plane $H_{\mathbb{C}a}$ by a discrete group Γ of isometries of the corresponding hyperbolic space H. These spaces again have negative curvature. In fact, the sectional curvature is $\equiv -1$ for quotients of $H_{\mathbb{R}}$ and lies between -4 and -1 in the other cases (for some suitable normalization). For quotients of $H_{\mathbb{Q}}$ and $H_{\mathbb{C}a}$, b_1 vanishes. There exist quotients of $H_{\mathbb{C}}$ with vanishing b_1, but there also exist such quotients with nonvanishing b_1.

---

[2]See the appendix to this §.

By Mostow's rigidity theorem, such spaces cannot be deformed inside the class of locally symmetric spaces, i.e. if we have two such spaces N_1, N_2 and an isomorphism

$$\alpha : \pi_1(N_1) \to \pi_1(N_2),$$

then α is induced by an isometry between the universal covers of N_1 and N_2. In particular, the mapping class group is finite, since any diffeomorphism $\psi : N \to N$ is homotopic to an isometry, and any compact manifold of negative sectional curvature admits at most finitely many isometries (actually, negative Ricci curvature already suffices for this property).

Again, any selfmap $\varphi : N \to N$ satisfies

$$|\deg \varphi| \leq 1.$$

There may still be infinitely many different homotopy classes of selfmaps. Namely, if $b_1(N) \neq 0$, then one may construct a continuous homotopically nontrivial map $\sigma : N \to S^1$, and mapping S^1 into N as a nontrivial closed geodesic, $\gamma : S^1 \to N$, the maps

$$\gamma \circ t^\nu \circ \sigma : N \to N$$

are mutually homotopic for different $\nu \in \mathbb{Z}$ where $t^\nu : S^1 \to S^1$ is a covering of degree ν. Since N has strictly negative curvature, its Riemannian metric can be deformed in the class of negative curvature metrics.

4) Compact irreducible locally symmetric spaces N of noncompact type and rank ≥ 2. Such spaces have nonpositive sectional curvature, but the curvature is not strictly negative anymore.

By a theorem of Matsushima [Mt]

$$b_1(N) = 0.$$

Mostow's rigidity theorem again applies.

Again, for any selfmap $\varphi : N \to N$

$$|\deg \varphi| \leq 1,$$

and in fact much more is true: If $\deg \varphi = \pm 1$, then φ is homotopic to an isometry, whereas if $\deg \varphi = 0$, then φ is homotopic to a constant map. Since the isometry group is again finite, it follows that the group of homotopy classes of selfmaps is finite as well. This result is a consequence of Margulis superrigidity theory.

Finally, by a theorem of Gromov, N is even rigid in the class of metrics of nonpositive sectional curvature. One should note that N contains totally geodesic embedded flat tori T^m of dimension $m = \mathrm{rank} N\, (> 1)$, but the topological and rigidity properties are characteristically different from, even typically opposite to those of a torus. Ironically, it is even this system of totally geodesic flat tori in N that makes N so rigid.

There are surprisingly few known examples of manifolds of negative or nonpositive curvature besides those just discussed. Let us briefly describe some general constructions:

Gromov and Thurston [GT] obtained examples of negatively curved compact Riemannian manifolds of dimension ≥ 3 that are not homeomorphic to locally hyperbolic spaces. They started with a compact quotient N of real hyperbolic space containing a compact totally geodesic submanifold D of codimension 2. One may then form a cyclic cover N' of N that is branched along D. The induced metric on N' then becomes singular along the preimage D' of D, but is otherwise smooth and of negative curvature. By performing some geometric surgery near D', they were then able to construct a smooth metric of negative curvature on N'. Given $\delta > 0$, one can construct examples with curvature between $-1 - \delta$ and -1.
A related construction had been used earlier in complex geometry by Mostow-Siu [MS]. They started with a compact quotient N of complex hyperbolic space of complex dimension 2 with a totally geodesic smooth divisor D, i.e. compact holomorphic curve ("curve" here refers to the complex dimension which is 1; thus the real dimension and the real codimension are both 2), and they constructed a negatively curved Kähler metric on some cover N' of N branched along D. This kind of construction was generalized by Zheng [Zh1] who was also able to handle the much more difficult case where D has singularities, although in the presence of singularities, he constructed only Kähler metrics of nonpositive, but not necessarily of negative curvature. See also [Zh2] for more examples in this direction.

In this context, we also draw attention to another construction developed by Abresch-Schroeder that produces metrics of nonpositive curvature. Also, in that direction, Hummel-Schroeder [HS] investigated conditions under which cusps of pinched negative curvature can be closed as manifolds with nonpositive sectional curvature, and in particular obtained new explicit examples of such manifolds.
Another method, due to Heintze [Hn] and Gromov [G1], produces metrics of nonpositive curvature by surgery along totally geodesic tori of codimension 1. For example, one may find noncompact finite volume quotients N of three dimensional real hyperbolic space whose fundamental group contains $\mathbb{Z}^2$ as a subgroup (note that in the compact case this is excluded by Preissmann's theorem). Geometrically, this corresponds to an injection $i : T^2 \to N$ ($T^2 =$ two-dimensional torus) with injective induced map $i_\sharp : \pi_1(T^2) \to \pi_1(N)$ on the level of fundamental groups. However, for any $\alpha \in \pi_1(T^2)$, $i_\sharp(\alpha)$, although not homotopic to a constant curve, can be deformed into an arbitrarily short loop by moving sufficiently far into a so-called cusp. Let us visualize this by the analogous picture in dimension 2, where the torus becomes one-dimensional, i.e. S^1:

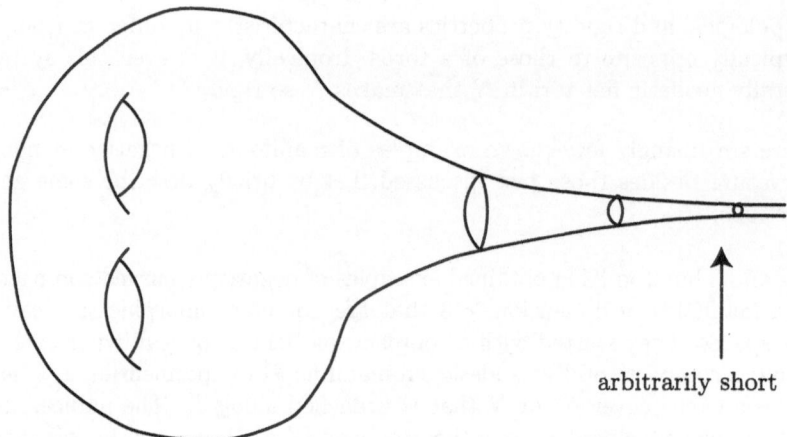

arbitrarily short

One may now perturb the metric in a neighborhood of this cusp so that its curvature stays nonpositive, but a suitable embedding $i_1(T^2)$ of T^2 becomes totally geodesic. Again, we have a picture in dimension 2:

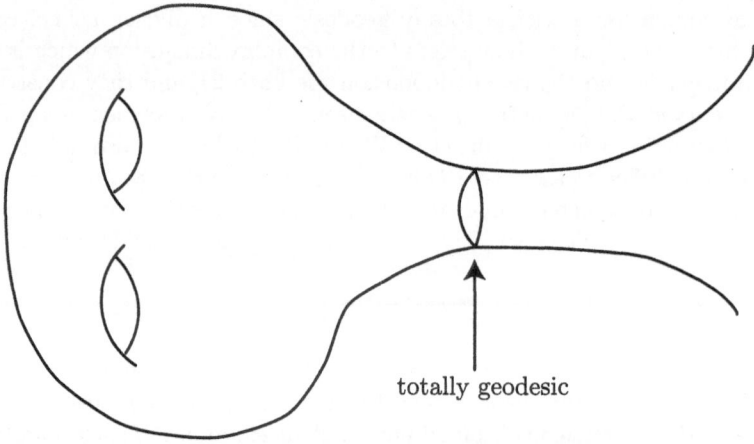

totally geodesic

We may now chop off the cusp outside $i_1(T^2)$. The remaining portion of N with its nonpositive curvature metric just described is called N_1. We may then take another space N_2 obtained by a similar kind of construction, with a totally geodesic embedded torus $i_2(T^2)$ that is isometric to $i_1(T^2)$, and glue N_1 and N_2 together by identifying $i_1(T^2)$ and $i_2(T^2)$ via an isometry.

The result is a compact manifold $\overline{N}$ of nonpositive curvature. It cannot admit a metric of negative curvature, by Preissmann's theorem again, since $\mathbb{Z}^2$ is a subgroup of its fundamental group. Nevertheless, in several geometric aspects, $\overline{N}$ is similar to a manifold of negative curvature, rather than for example to a locally symmetric space of higher rank (see [Ba]).

Besides Preissmann's theorem and its extensions (see below), there is another restriction for a group Γ to be the fundamental group of a compact manifold of negative curvature, found by Švarc [Sv] and Milnor [Mi]. Namely, Γ has to be of exponential growth. This means the following: Let $\{\gamma_1, \ldots, \gamma_m\}$ be a set of generators of Γ. Thus, every element of Γ is a product of elements from the set $S_\Gamma := \{\gamma_1, \gamma_1^{-1}, \gamma_2, \gamma_2^{-1}, \ldots, \gamma_m, \gamma_m^{-1}\}$, like $\gamma_1 \gamma_2^2 \gamma_1^{-1} \gamma_2^3 \gamma_1$ (note that any element of S_Γ may occur repeatedly). Let Γ_n be the set of elements of Γ that can be represented by products of at most n elements from S_Γ, counted with repetition. That Γ has exponential growth then means that

$$\sharp \Gamma_n \geq c_1 e^{c_2 n} \quad \text{for some positive constants } c_1, c_2.$$

By way of contrast, the fundamental group of a compact manifold of nonnegative Ricci curvature, for example of a torus, always has polynomial growth, i.e.

$$\sharp \Gamma_n \leq c n^k \quad \text{for some } c > 0, \ k \in \mathbb{N}.$$

Farrell-Jones [FJ2] started with a compact quotient N of $H_\mathbb{R}^n$ for $n \geq 5$ and formed connected sums $\widehat{N}$ of N with exotic spheres. They showed that such spaces $\widehat{N}$ while not diffeomorphic to N are homeomorphic to N and carry metrics of negative sectional curvature. Given $\delta > 0$, one may produce examples with curvature between $-1 - \delta$ and -1. Likewise, for compact quotients N of $H_\mathbb{C}^n$, $n = 4$ or $n = 4m + 1$ for some $m \in \mathbb{N}$, they formed such connected sums $\widehat{N}$ of N with suitable exotic spheres which again are homeomorphic, but not diffeomorphic to N and carry metrics of negative curvature. Given $\delta > 0$, one may now find examples with curvature betweeen $-4 - \delta$ and -1 (note that N has curvature between -4 and -1). See [FJ4].

In [FJ1], they showed the following topological analogue of Mostow's rigidity theorem: A compact manifold N of dimension $n \geq 5$ carries a (real) hyperbolic structure (i.e. a metric with sectional curvature -1) if and only if $\pi_1(N)$ is isomorphic to a discrete cocompact subgroup of $O(n,1)$ (the isometry group of $H_\mathbb{R}^n$) and $\pi_k(N) = \{1\}$ for all $k \geq 2$.

One may ask certain questions that, to the author's knowledge, so far have not been successfully considered:

1) Characterize those discrete groups that occur as fundamental groups of compact manifolds of negative or nonpositive curvature, i.e. find necessary and sufficient conditions for a group to be the fundamental group of such a space. Of course, in the negative curvature case, some necessary conditions are provided by the theorems of Preissmann and Švarc-Milnor. Also, because of Corollary 1.1.1, the cohomological dimension of such a group Γ has to be equal to the dimension of the compact manifold N of nonpositive curvature whose fundamental group it is. (By Corollary 1.1.1, N is a so-called $K(\Gamma, 1)$ space, meaning that the cohomology of Γ coincides with that of N.) However, all these conditions seem very far from being sufficient.

Probably, the question is not even well posed in its present form. Namely, in Chapter 2, we shall see natural generalizations of the concept of a manifold of nonpositive sectional curvature to which the previous results still largely apply. Therefore, one should rather ask for a characterization of fundamental groups of compact such generalized spaces of nonpositive or negative curvature. Still, this may not be a wellposed question. Burger-Mozes [BM] constructed for all sufficiently large integers (m, n) a finite complex with a *simple* fundamental group whose universal covering is the product of regular trees of resp. degrees m and n.

Gromov [G3] pointed out that fundamental groups of compact spaces of negative curvature satisfy the algebraic property of being word hyperbolic. To explain this notion, let Γ be a finitely generated group. Fix a finite system of generators of Γ, and denote the length of the shortest word representing $\gamma \in \Gamma$ in these generators by $|\gamma|$ (i.e. if we write γ as a product of these generators and their inverses, then $|\gamma|$ is the smallest possible number of factors in such a product). Gromov puts

$$\alpha \cdot \beta := \frac{1}{2}\left(|\alpha| + |\beta| - |\alpha^{-1}\beta|\right) \quad \text{for } \alpha, \beta \in \Gamma$$

and calls Γ word hyperbolic if there exists $\delta \geq 0$ (that will depend on the choice of the generators) with

$$\alpha \cdot \beta \geq \min(\alpha \cdot \gamma, \beta \cdot \gamma) - \delta \quad \text{for all } \alpha, \beta, \gamma \in \Gamma.$$

An equivalent characterization is the following isoperimetric inequality. For every finitely presented group Γ, one may construct a connected bounded domain $\Omega \subseteq \mathbb{R}^5$ with a smooth boundary with fundamental group Γ. Γ is word hyperbolic if there exists a constant C (that will depend on Ω) with the property that every smooth simple closed contractible curve S in Ω bounds a smooth embedded disk D in Ω with

$$\text{Area}(D) \leq C \, \text{length}(S).$$

Again, this type of isoperimetric inequality holds of course in the universal cover of a compact space of negative curvature. One might then ask for a characterization of word hyperbolic groups. In this context, Olshanski [Ol] showed that any nonelementary word hyperbolic group Γ is SQ-universal, meaning that any countable group can be imbedded in a quotient of Γ.

One does not know wether the question if a finitely presented group Γ is word hyperbolic is decidable. However, Sela [Sl] (elaborating on work of Rips) showed that the isomorphism problem for word hyperbolic groups is solvable. (The preceding comments and references were pointed out to me by Marc Burger.)

2) The mapping class group, i.e. the group of homotopy classes of diffeomorphisms, is finite for a compact manifold of negative curvature and dimension ≥ 3 by a theorem of Gromov. Can one give explicit estimates for its order in terms of topological data?

3) For a compact hyperbolic Riemann surface of genus $g \geq 2$ the number of isometries is bounded by $84(g-1)$ (Theorem of Hurwitz). In fact, this holds for any metric on such a surface, not necessarily of constant negative curvature, since an isometry is always conformal for the underlying Riemann surface structure and therefore also an isometry w.r.t. the hyperbolic metric.

Can one estimate the number of isometries of a compact manifold of negative curvature and dimension ≥ 3 in terms of natural geometric data? Some such estimate was obtained by Im Hof [IH] in terms of upper and lower curvature bounds, dimension, diameter, and injectivity radius. The estimate does not seem to be sharp, however. Also, it was shown by X.Z. Dai - Z.M. Shen - G.F. Wei [DSW], that the number of isometries for a manifold of negative Ricci curvature, the order of the isometry group can be estimated in terms of upper and lower Ricci curvature bounds, dimension, a lower injectivity radius bound and an upper volume bound. This is a quantitative version of Bochner's theorem on the finiteness of the isometry group in that class of manifolds. The proof, however, is indirect, and no explicit estimate is obtained. [Sl] has a result characterizing the word hyperbolic groups with finite outer automorphism group (again pointed out to me by Marc Burger).

Appendix to § 1.1: Symmetric spaces of noncompact type

In order to better understand the preceding examples and many points in the sequel, it is useful to recall the notion and structure of symmetric spaces of noncompact type. A symmetric space of noncompact type can be realized as

$$M = G/K$$

where G is a noncompact semisimple Lie group with trivial center, and K is a maximal compact subgroup. M is a differentiable manifold, and it carries a Riemannian metric of nonpositive curvature for which G acts by isometries.
A typical example is

$$M = Sl(n, \mathbb{R})/SO(n),$$

and we shall now exhibit some typical features of the geometry of symmetric spaces of noncompact type at this example.

Some notation:

$$Gl(n,\mathbb{R}) \quad := \quad \{(n \times n)\,\text{matrices } A \text{ over } \mathbb{R} \text{ with det } A \neq 0\}$$
$$\text{(general linear group)}$$
$$Sl(n,\mathbb{R}) \quad := \quad \{A \in Gl(n,\mathbb{R}) : \det A = 1\}\,\text{(special linear group)}$$
$$SO(n) \quad := \quad \{A \in Sl(n,\mathbb{R})) : A^t = A^{-1}\}\,\text{(special orthogonal group)}$$

These are Lie groups with associated Lie algebras

$$\mathfrak{gl}(n,\mathbb{R}) \quad := \quad \{(n \times n)\,\text{matrices } A \text{ over } \mathbb{R}\}$$
$$\mathfrak{sl}(n,\mathbb{R}) \quad := \quad \{X \in \mathfrak{gl}(n,\mathbb{R}) : \operatorname{tr} X = 0\}$$
$$\mathfrak{so}(n) \quad := \quad \{X \in \mathfrak{sl}(n,\mathbb{R})) : X^t = -X\}$$

(We always denote the Lie algebra of a Lie group G by the corresponding gothic letter $\mathfrak{g}$.)

The exponential map yields a diffeomorphism between a neighborhood of O in a Lie algebra $\mathfrak{g}$ and a neighborhood of the identity in the corresponding Lie group G. In the present case where $G \subset Gl(n,\mathbb{R})$, the exponential map is the usual matrix exponential map:

For $X \in \mathfrak{gl}(n,\mathbb{R})$

$$\exp X := e^X := \operatorname{Id} + X + \frac{1}{2}X^2 + \ldots \tag{1.1.1}$$

For $X, Y \in \mathfrak{gl}(n,\mathbb{R})$, we put

$$(\operatorname{ad}X)Y := [X,Y] := XY - YX. \tag{1.1.2}$$

The first part of of this formula also holds in general for $X \in \mathfrak{g}$, a Lie algebra, if $[\cdot,\cdot]$ denotes the Lie bracket, and ad is the derivative of conjugation by an element of G.

For a Lie algebra $\mathfrak{g}$, we define the Killing form as

$$B(X,Y) = \operatorname{tr} \operatorname{ad}X \operatorname{ad}Y \quad \text{for } X, Y \in \mathfrak{g} \tag{1.1.3}$$

where $\operatorname{ad}X$, $\operatorname{ad}Y$ are considered as linear endomorphisms of $\mathfrak{g}$. $\mathfrak{g}$ and G are called semisimple if B is nondegenerate.

On $\mathfrak{gl}(n,\mathbb{R})$, we have

$$B(X,Y) = \operatorname{tr} \operatorname{ad}X \operatorname{ad}Y = 2n\operatorname{tr}XY - 2\operatorname{tr}X\operatorname{tr}Y$$

where the trace on the r.h.s. is the usual matrix trace. For $X = \alpha\operatorname{Id}$, we have

$$\operatorname{ad}X = 0$$

and therefore $\mathfrak{gl}(n,\mathbb{R})$ is not semisimple. $\mathfrak{sl}(n,\mathbb{R})$, however, is semisimple, and we have for $X, Y \in \mathfrak{sl}(n,\mathbb{R})$

$$B(X,Y) = 2n\operatorname{tr}XY, \tag{1.1.4}$$

in particular

$$B(X, X^t) > 0 \quad \text{for } X \neq 0, \tag{1.1.5}$$

showing nondegeneracy.

We now return to $M = Sl(n, \mathbb{R})/SO(n)$. To simplify notation, we put $G = Sl(n, \mathbb{R})$, $K = SO(n)$, and $\mathfrak{g} = \mathfrak{sl}(n, \mathbb{R})$, $\mathfrak{k} = \mathfrak{so}(n)$. Any $A \in GL(n, \mathbb{R})$ can be uniquely decomposed (Cauchy polar decomposition theorem) as

$$A = VR \tag{1.1.6}$$

with a symmetric positive definite matrix V and an orthogonal matrix R. We put

$$P := \{A \in Sl(n, \mathbb{R}) : A^t = A, A \text{ positive definite}\}$$
$$\mathfrak{p} := \{X \in \mathfrak{sl}(n, \mathbb{R}) : X^t = X\}$$

(i.e. $\mathfrak{g} = \mathfrak{k} \oplus \mathfrak{p}$, orthogonal decomposition w.r.t B, and $V \in P$ in the above decomposition). In fact, one deduces from the polar decomposition theorem that $M = Sl(n, \mathbb{R})/SO(n)$ is homeomorphic to P, and since P is naturally a differentiable manifold, so then is M. G operates transitively on G/K by diffeomorphisms (here $G = Sl(n, \mathbb{R})$, $K = SO(n)$, but this holds in general, like all the structural results that we shall describe):

$$G \times G/K \rightarrow G/K$$
$$(h, gK) \rightarrow hgK.$$

The isotropy group of $\text{Id} \cdot K$ is K, while the isotropy group of gK is gKg^{-1} which is conjugate to K.

In order to describe the metric, we put

$$\langle X, Y \rangle_{\mathfrak{g}} := \begin{cases} B(X, Y) & \text{for } X, Y \in \mathfrak{p} \\ -B(X, Y) & \text{for } X, Y \in \mathfrak{k} \\ 0 & \text{otherwise.} \end{cases} \tag{1.1.7}$$

It follows from (1.1.4) that $\langle \cdot, \cdot \rangle_{\mathfrak{g}}$ is positive definite.

We put $e := \text{Id} \in G$, and we identify $T_e G$ with $\mathfrak{g}$. $\langle \cdot, \cdot \rangle_{\mathfrak{g}}$ then defines a metric on $T_e G$, and we then obtain a metric on $T_g G$ for arbitrary g by requiring that the left translation

$$L_g : G \rightarrow G$$
$$h \rightarrow gh$$

induces an isometry between $T_e G$ and $T_g G$. Likewise, we get an induced metric on G/K: restricting $\langle \cdot, \cdot \rangle_{\mathfrak{g}}$ to $\mathfrak{p}$ yields a metric on $T_{eK} G/K \simeq \mathfrak{p}$, and we require again that

$$L_g : G/K \rightarrow G/K$$
$$hK \rightarrow ghK$$

induces an isometry between $T_{eK}G/K$ and $T_{gK}G/K$. One checks that this metric is well defined, and G then operates isometrically on G/K.

In order to show that G/K is a symmetric space in the usual differential geometric sense, we need to describe an appropriate involution. We start with

$$\begin{aligned} \tau_e : G &\rightarrow G \\ h &\rightarrow (h^{-1})^t, \quad \text{in particular } \tau_{e|K} = \text{id} \end{aligned}$$

with derivative

$$\begin{aligned} d\tau_e : \mathfrak{g} &\rightarrow \mathfrak{g} \\ X &\rightarrow -X^t. \end{aligned}$$

Therefore,

$$d\tau_{e|\mathfrak{k}} = \text{id}, \quad d\tau_{e|\mathfrak{p}} = -\text{id}.$$

Next, for $g \in G$

$$\begin{aligned} \tau_g : G &\rightarrow G \\ h &\rightarrow gg^t(h^{-1})^t. \end{aligned}$$

Then

$$\tau_g^2 = \text{Id}, \quad \text{and } \tau_g(g) = g.$$

Consequently, we get corresponding involutions

$$\tau_{gK} : G/K \rightarrow G/K$$

with

$$\tau_{gK}(gK) = gK, \quad d\tau_{gK} = -\text{id}, \quad \tau_{gK}^2 = \text{id}, \tag{1.1.8}$$

and G/K is a symmetric space, indeed.

Moreover, the matrix exponential map (1.1.1) when restricted to $\mathfrak{p}$ becomes the Riemannian exponential map for G/K at $T_{eK}G/K \simeq \mathfrak{p}$.

Since G operates by isometries on G/K, each $X \in \mathfrak{g} \simeq T_e G$ becomes a Killing vector field on G/K. Since the stabilizer of $eK \in G/K$ is in K, the elements of $\mathfrak{k}$ represent those Killing fields that vanish at eK. Also, for $Y \in \mathfrak{p}$, the geodesic

$$c(t) = \exp_{eK} tY(eK)$$

satisfies

$$Y(c(t)) = \dot{c}(t) \quad \text{for all } t \in \mathbb{R}.$$

Since any Killing field is a Jacobi field along any geodesic, we get the Jacobi equation

$$\nabla_Y \nabla_Y X + R(X, Y)Y = 0$$

along c, hence in particular at eK. Here, ∇ denotes the Levi-Civita connection of G/K.

For $Y, Z \in \mathfrak{p}$, we also have $Y + Z \in \mathfrak{p}$, and we deduce

$$\nabla_Y \nabla_Z X + \nabla_Z \nabla_Y X + R(X,Y)Z + R(X,Z)Y = 0.$$

Now

$$R(Y,Z)X = \nabla_Y \nabla_Z X - \nabla_Z \nabla_Y X - \nabla_{[Y,Z]} X$$
$$R(X,Z)Y = -R(Z,X)Y$$
$$R(X,Y)Z + R(Y,Z)X + R(Z,X)Y = 0 \quad \text{(Bianchi identity)},$$

and since $[Y,Z] \in \mathfrak{k}$ for $Y, Z \in \mathfrak{p}$, we also have

$$[Y,Z](eK) = 0, \tag{1.1.9}$$

since isometries induced by $k \in K$ leave eK fixed.
Altogether, we obtain

$$\nabla_Y \nabla_Z X + R(X,Y)Z = 0. \tag{1.1.10}$$

We obtain at eK

$$
\begin{aligned}
R(X,Y)Z &= -R(Y,Z)X + R(X,Z)Y \quad \text{(Bianchi identity)} \\
&= \nabla_Z \nabla_X Y - \nabla_Z \nabla_Y X \quad \text{(by (1.1.10))} \\
&= \nabla_Z [X,Y] \quad \text{(since } \nabla \text{ is torsionfree)} \\
&= \nabla_{[X,Y]} Z - [[X,Y],Z] \quad \text{(again, since } \nabla \text{ is torsionfree)} \\
&= -[[X,Y],Z] \quad \text{(since } [X,Y](e,K) = 0 \text{ by (1.1.9)).}
\end{aligned}
$$

By applying left translations, we then get this formula at every $p \in G/K$, i.e.

Theorem 1.1.3: *With the identification* $T_p G/K \simeq \mathfrak{p}$, *the curvature tensor of* G/K *satisfies*

$$R(X,Y)Z = -[[X,Y],Z] \quad \text{for } X, Y, Z \in \mathfrak{p}. \tag{1.1.11}$$

In particular, the sectional curvature of a plane in $T_p M$ spanned by orthonormal vectors Y_1, Y_2 is given by

$$
\begin{aligned}
K(Y_1 \wedge Y_2) &= -\langle [[Y_1,Y_2],Y_2],Y_1 \rangle \\
&= -B([[Y_1,Y_2],Y_2],Y_1) \\
&= -B([Y_2,[Y_2,Y_1]],Y_1) \\
&= B([Y_2,Y_1],[Y_2,Y_1]) \quad \text{because of the} \tag{1.1.12}
\end{aligned}
$$

Ad invariance of the Killing
form (we simply have
$\operatorname{tr}[Z,X]Y + \operatorname{tr}X[Z,Y] = 0$).

We have the general relations

$$[\mathfrak{k}, \mathfrak{k}] \subset \mathfrak{k}$$
$$[\mathfrak{p}, \mathfrak{p}] \subset \mathfrak{k} \quad \text{(in particular, } \mathfrak{p} \text{ is \underline{not} a Lie algebra)}$$
$$[\mathfrak{p}, \mathfrak{k}] \subset \mathfrak{p}.$$

Therefore, for $Y_1, Y_2 \in \mathfrak{p}$, $[Y_1, Y_2] \in \mathfrak{k}$, and since the Killing form is negative definite on $\mathfrak{k}$ (by (1.1.5) since $X^t = -X$ for $X \in \mathfrak{k}$), we obtain from (1.1.12)

Corollary 1.1.2: *The symmetric space* $Sl(n, \mathbb{R})\big/SO(n)$ *has nonpositive sectional curvature.*

This holds for all symmetric spaces of noncompact type, by a generalization of the preceding arguments.

From the proof of Theorem 1.1.3 we see that $K(Y_1 \wedge Y_2)$ is negative unless $[Y_1, Y_2] = 0$, in which case $K(Y_1 \wedge Y_2) = 0$.
Thus, in particular the sectional curvature is negative if any two linearly independent Y_1, Y_2 have nonvanishing Lie bracket.

Definition 1.1.2: A Lie subalgebra $\mathfrak{a}$ of $\mathfrak{g}$ is called *Abelian* if $[A_1, A_2] = 0$ for all A_1, $A_2 \in \mathfrak{a}$.

Let $\mathfrak{a}$ be an Abelian subalgebra of $\mathfrak{g}$ contained in $\mathfrak{p}$. This means that the elements of $\mathfrak{a}$ constitute a commuting family of symmetric matrices. Therefore, they can be diagonalized simultaneously, and there exists an orthonormal basis $v_1, \ldots, v_n$ of $\mathbb{R}^n$ of common eigenvectors of the elements of $\mathfrak{a}$.
Let $e_1, \ldots, e_n$ be some fixed orthonormal basis of $\mathbb{R}^n$. We may then find $R \in SO(n)$ with

$$R(v_j) = e_j \quad \text{for } j = 1, \ldots, n.$$

Therefore, $R\mathfrak{a}R^{-1}$ is an Abelian subspace of $\mathfrak{p}$ with eigenvectors $e_1, \ldots, e_n$, and with respect to this fixed basis of $\mathbb{R}^n$, the elements of $R\mathfrak{a}R^{-1}$ are diagonal matrices (tracefree since contained in $\mathfrak{p}$). We conclude that the space of tracefree diagonal matrices is a maximal Abelian subspace of $\mathfrak{p}$. Each maximal Abelian subspace of $\mathfrak{p}$ is conjugate to this special one, w.r.t. some $R \in SO(n)$. In particular, any two maximal subspaces of $\mathfrak{p}$ are conjugate.

For a (not necessarily maximal) Abelian subspace $\mathfrak{a}$ of $\mathfrak{p}$, we define

$$A := \exp \mathfrak{a}$$

which is an Abelian Lie subgroup of G, since

$$g_1 g_2 = g_2 g_1 \quad \text{for all } g_1, g_2 \in A.$$

Since $\mathfrak{a} \subset \mathfrak{p}$, A can also be considered as a subspace of $G\big/K$.

Theorem 1.1.4: *A is totally geodesic and flat in M ("flat" means that the curvature is $\equiv 0$).*

That A is totally geodesic follows from the subsequent lemma, and the flatness is a consequence of (1.1.13). $\square$

Lemma 1.1.2: *Let $\mathfrak{p}^*$ be a subspace of $\mathfrak{p}$ with $[X,[Y,Z]] \in \mathfrak{p}^*$ for all X, Y, $Z \in \mathfrak{p}^*$ (such a $\mathfrak{p}^*$ is called a Lie triple system).*
Then $M^* := \exp_{eK}(\mathfrak{p}^*)$ is a totally geodesic submanifold of M.

Proof: We put $\mathfrak{k}^* := [\mathfrak{p}^*,\mathfrak{p}^*]$, $\mathfrak{g}^* = \mathfrak{k}^* \oplus \mathfrak{p}^*$. $\mathfrak{g}^*$ then is a Lie subalgebra of $\mathfrak{g}$, and $[\mathfrak{k}^*,\mathfrak{k}^*] \subset \mathfrak{k}^*$, $[\mathfrak{p}^*,\mathfrak{p}^*] \subset \mathfrak{k}^*$, $[\mathfrak{k}^*,\mathfrak{p}^*] \subset \mathfrak{p}^*$, because $\mathfrak{p}^*$ is a Lie triple system.
G^*/H^* ($G^* :=$ connected Lie subgroup of G with Lie algebra $\mathfrak{g}^*$, $H^* := \{g \in G^* : gK = K\}$) then becomes a submanifold of G/K.
Since the exponential map of $\mathfrak{g}^*$ is the restriction of the exponential map of $\mathfrak{g}$, we see that any geodesic that is tangent to G^*/H^* at some point stays in G^*/H^* which is therefore totally geodesic. $\square$

It is also not hard to verify the converse of Lemma 1.1.2, namely that every complete totally geodesic subspace of G/K comes from a Lie triple system. In fact, one has

Theorem 1.1.5: *Every symmetric space of noncompact type which is irreducible (i.e. not a nontrivial product of such spaces) can be isometrically (up to scaling the metric by a constant factor) embedded as a complete, totally geodesic submanifold of $Sl(n,\mathbb{R})/SO(n)$.*

We omit the proof although it is not hard (see e.g. [Eb], pp. 134ff.) and return to the flat subspaces of Theorem 1.1.4. That theorem, and the discussion preceding it, imply that the maximal flat subspaces of G/K through eK bijectively correspond to the maximal Abelian subspaces of $\mathfrak{p}$. ("Maximal" meaning "not contained in a larger one".)

Definition 1.1.3: The *rank* of a symmetric space is the dimension of a maximal flat subspace (or, equivalently, the dimension of a maximal Abelian subspace of $\mathfrak{p}$).

Since the maximal Abelian subspace of $\mathfrak{p}$ for $\mathfrak{g} = \mathfrak{sl}(n,\mathbb{R})$ consists of the tracefree diagonal matrices, we get

$$\text{Rank}\left(Sl(n,\mathbb{R})/SO(n)\right) = n - 1.$$

Irreducible (i.e. not being decomposable as a nontrivial product) symmetric spaces can be classified into a finite number of series like $Sl(n,\mathbb{R})/SO(n)$ for $n \in \mathbb{N}$ plus a finite number of exceptional ones. (Since every symmetric space is a product of irreducible ones, this provides a complete classification.) See Helgason [He].
The symmetric spaces of noncompact type and rank 1 are found to be the hyperbolic spaces over $\mathbb{R}$, $\mathbb{C}$, $\mathbb{Q}$ (quaternions), denoted by $H_{\mathbb{R}}^n$ (dim $= n$), $H_{\mathbb{C}}^n$ (dim $= 2n$),

$H_{\mathbb{Q}}^n$ (dim $= 4n$), plus the hyperbolic Cayley plane $H_{\mathbb{C}a}^2$ (dim $= 16$). We have

$$
\begin{aligned}
H_{\mathbb{R}}^n &= SO_o(n,1)/SO(n) \\
H_{\mathbb{C}}^n &= SU(n,1)/S(U(n) \times U(1)) \\
H_{\mathbb{Q}}^n &= Sp(n,1)/Sp(n) \times Sp(1) \\
H_{\mathbb{C}a}^2 &= F_4^{-20}/Spin(9).
\end{aligned}
$$

By a theorem of Borel, every symmetric space G/K of noncompact type admits a smooth compact quotient. This means that one can always find a discrete, torsion-free, cocompact subgroup Γ of G. "Discrete" means that Γ has no accumulation points w.r.t. the topology of the Lie group G, "torsionfree" that Γ does not contain elements of finite order other than the identity (for the induced action of Γ on G/K this means that $\gamma p = p$ for some $p \in G/K$ implies $\gamma = $ id), and "cocompact" that $\Gamma \backslash G/K$ is compact. $\Gamma \backslash G/K$ then becomes a compact Riemannian manifold that is locally isometrically covered by G/K. In particular, $\Gamma \backslash G/K$ has nonpositive sectional curvature. Since the actions of K and Γ on G/K do not commute, G does not act anymore on $\Gamma \backslash G/K$. $\Gamma \backslash G/K$ is a so-called locally symmetric space of noncompact type. A discrete cocompact subgroup Γ of G/K is called a uniform lattice.

There also exist discrete subgroups Γ of G for which $\Gamma \backslash G/K$ is not compact, but has finite volume for the induced metric (this is again a result of Borel). Such subgroups are called nonuniform lattices. Any lattice admits a torsionfree subgroup of finite index. Therefore, one finds examples both of compact and noncompact finite volume locally symmetric spaces of noncompact type.

References for a geometric treatment of symmetric spaces of noncompact type are

W. Ballmann, M. Gromov, V. Schroeder, Manifolds of nonpositive curvature, Birkhäuser, 1985

P. Eberlein, Structure of manifolds of nonpositive curvature, Springer LNM 1156, 1984, pp. 86–153

J. Jost, Riemannian geometry and geometric analysis, Springer, 1995

1.2 Mordell and Shafarevitch type problems over function fields: An example of negative curvature in algebraic geometry

In this §, we shall try to describe some of the geometric content of the following theorems.

Theorem 1.2.1: *Let C be a compact smooth holomorphic curve (= a compact Riemann surface in different terminology), and let $g \in \mathbb{N}$, $g \geq 2$. Then there exist at most finitely many algebraic surfaces B ("surface" here means complex dimension 2) fibered over C with smooth fibres of genus g in a nontrivial manner. (This means that there exists a holomorphic map*

$$f : B \to C$$

for which $f^{-1}(z)$ is a smooth holomorphic curve of genus g for every $z \in C$. "Nontrivial" means that one cannot find finite covers B' of B, C' of C with $B' = \Sigma \times C'$ for some holomorphic curve Σ. More precisely, one calls a fibering that is covered by a product "isotrivial".)

This result is due to Parshin [Pa]. It was generalized by Arakelov to the case where one has a finite subset S of C and a fibering $f : B \to C$ for which the fibers over $C \setminus S$ only are required to be smooth holomorphic curves of genus g, while the fibers over S are allowed to have singularities. This represents the solution of the so-called Shafarevitch problem over function fields.

Theorem 1.2.2: *Let $f : B \to C$ be a nontrivial fibering as described in Theorem 1.2.1, again with fibers of genus $g \geq 2$. Such a fibering admits only finitely many holomorphic sections $s : C \to B$ (that s is a section means that $f \circ s = id$).*

This result was first obtained by Manin [Ma], and Grauert [Gr] found a different proof shortly afterwards. Theorem 1.2.2 represents the solution of the so-called Mordell problem over function fields.

For our discussion of these results, we need the moduli space M_g of holomorphic curves of genus g (≥ 2). It is the space of all holomorphic, or equivalently, conformal structures on a given topological surface of genus g, equipped with some natural topology. It is not compact, but there exist natural compactifications that can be given the structure of an algebraic variety, for example the Mumford-Deligne compactification $\overline{M_g}$. M_g is not a manifold, because it has certain singularities caused by those holomorphic curves that admit nontrivial automorphisms (= bijective holomorphic selfmaps). Some finite cover M'_g, of M_g, however is a manifold. M_g has the following universal property: Whenever

$$f : B \to C$$

is a fibering by curves of genus g as in Theorem 1.2.1, we obtain a holomorphic map

$$h : C \to M_g$$

by associating to $z \in C$ the holomorphic structure of the curve $f^{-1}(z)$.
We then also obtain a holomorphic map between finite covers,

$$h' : C' \to M'_g.$$

h' is constant iff the fibering is isotrivial.

M'_g carries a natural Kähler metric, the so-called Weil-Petersson metric g_{WP}. g_{WP} is not complete, since $\partial \overline{M'_g} := \overline{M'_g} \setminus M_g$ has finite distance from the interior. It was shown by Tromba [Tr] that g_{WP} has negative sectional curvature, and its holomorphic sectional curvature even has a negative upper bound $k < 0$. Different proofs of these results were found by Wolpert [Wo], Siu [Si3], Jost [Jo2], Jost-Peng [JP].

C' may be equipped with a metric of constant curvature $\kappa = 1$ if its genus is 0, of curvature $\kappa = 0$ if its genus is 1, and of curvature $\kappa = -1$ if its genus is ≥ 2. We may then apply the Schwarz type Lemma 1.1.1 to obtain

$$\|dh'(z)\|^2 \leq \frac{\kappa}{k} \quad \text{for all } z \in C', \tag{1.2.1}$$

unless h' is constant.

In particular, we obtain

Theorem 1.2.3: *If the genus of C is 0, 1, then any fibering $f : B \to C$ by curves of genus ≥ 2 is isotrivial.*

Proof: If genus$(C) = 0$, then $C = C'$ because S^2 does not admit unbranched coverings, while if genus$(C) = 1$, then any unbranched covering C' of C also has genus 1. Therefore,

$$\kappa \geq 0 \quad \text{if genus}(C) \leq 1.$$

Since $k < 0$, (1.2.1) implies that h' is constant. Therefore, the fibering is isotrivial.
□

In general, if the genus(C) is at least 2, we conclude from (1.2.1), that all possible holomorphic maps $h' : C' \to M'_g$ are equicontinuous. In particular, only finitely many homotopy classes of maps from C' to M'_g can contain holomorphic maps. Namely, if

$$h'_n : C' \to M'_g$$

is a sequence of holomorphic maps, it contains a convergent subsequence since being equicontinuous (applying the Arzela-Ascoli theorem), and therefore, for sufficiently large n, all h'_n are homotopic to the limit map h'_0. (There is a key technical point here, because one has to exclude that the image of h'_0 meets $\partial M'_g$. We omit the details.)

This step, namely that only finitely many homotopy classes of maps from C' to M'_g can be holomorphically represented, is called the boundedness part of the proof of Theorem 1.2.1. The preceding argument was taken from Jost-Yau [JY3]. A related argument had been found by Noguchi [No] who had exploited the fact that M'_g is known to be hyperbolic in the sense of Kobayashi by a result of Royden. For holomorphic maps into such hyperbolic spaces, one has a Schwarz lemma by definition, and so one obtains boundedness in the same manner. Ultimately, this approach can be traced back to the work of Grauert-Reckziegel [GR].

The second step of the proof, the so-called finiteness part, then consists in showing that any nontrivial homotopy class of maps from C' to M'_g can contain at most one holomorphic map. In Jost-Yau [JY3], negativity properties of the curvature of M'_g are again used to show finiteness.

For the proof of Theorem 1.2.2, one uses the holomorphic fibering

$$\pi : \mathcal{M}'_g \to M'_g$$

where the fiber of $p \in M'_g$ is given by the holomorphic curve that is defined by p. $\mathcal{M}'_g$ again carries a Weil-Petersson metric with the same negativity properties as the one of M'_g.

A holomorphic section s as in Theorem 1.2.2 then induces a holomorphic map

$$k : C' \to \mathcal{M}'_g.$$

The boundedness proof then can be obtained as above. (The finiteness can also be concluded on the basis of the negativity of the sectional curvature, see [JY3] again.)

We note that it is not all essential for the previous arguments that the base is a curve C. For the boundedness part, we could take any compact Hermitian manifold as base, while for the finiteness one has to require that the base is a Kähler manifold.

A similar strategy also works for many other types of families of algebraic varieties, because in many typical cases, the corresponding moduli spaces carry metrics with suitable negative curvature properties. For example, the moduli space of principally polarized Abelian varieties (i.e. tori of the form $\mathbb{C}^{2n}/\Gamma$ with a lattice Γ (isomorphic to $\mathbb{Z}^{2n}$ as an abstract group, but not necessarily as a subgroup of $\mathbb{C}^{2n}$) which admit a holomorphic embedding into some complex projective space $\mathbb{CP}^N$, together with a choice of an ample line bundle, i.e. a Riemann form) is given by the locally symmetric space $S_n := Sp(2n, \mathbb{Z}) \backslash {}^{Sp(2n, \mathbb{R})} / U(n)$.

Here, in order to describe $Sp(2n, \mathbb{R})$, we introduce the $2n \times 2n$ real matrix

$$\Omega = \begin{pmatrix} 0 & -\text{Id} \\ \text{Id} & 0 \end{pmatrix}$$

where Id is the $n \times n$ identity matrix. We put

$$(v, w) := \langle \Omega v, w \rangle \quad \text{for } v,\, w \in \mathbb{R}^{2n}$$

where $\langle \cdot, \cdot \rangle$ is the Euclidean scalar product. $Sp(2n, \mathbb{R})$ then is the subgroup of $Gl(2n, \mathbb{R})$ consisting of all matrices A preserving $(\cdot, \cdot)$, i.e.

$$(Av, Aw) = (v, w) \quad \text{for all } v,\, w \in \mathbb{R}^{2n}.$$

$Sp(2n, \mathbb{Z})$ is the subgroup of $Sp(2n, \mathbb{R})$ consisting of matrices with integer entries. $U(n)$ of course is the subgroup of $Gl(n, \mathbb{C}) \subset Gl(2n, \mathbb{R})$ consisting of unitary matrices.

Again, S_n is not a manifold, because some nontrivial elements act on $Sp(2n, \mathbb{R})/U(n)$ with fixed points, but again some finite cover S'_n is a manifold.

$Sp(2n, \mathbb{R})/U(n)$ is a symmetric space of noncompact type and rank n. It therefore has nonpositive sectional curvature, but for $n \geq 2$ the sectional curvature is not strictly negative. It is also a Hermitian symmetric space, meaning that it carries a natural complex structure for which the symmetric metric becomes a Kähler metric. Since the flat tori causing the zeros of the sectional curvature turn out to be totally real, the holomorphic sectional curvature is negative, in fact ≤ -1. Therefore, one may show boundedness for families of Abelian varieties of complex dimension n fibering over some given Kähler manifold by using the Schwarz type lemma 1.1.1 as above. Finiteness, however, is not true in general for families of Abelian varieties. One may simply take any such family and take its Cartesian product with any fixed Abelian variety T. This produces another family which is nontrivial if the original one was, but which can now be deformed by deforming T. Nevertheless, if one imposes a certain additional condition excluding this type of families, one may still show a finiteness result. This is due to Faltings [Fa]. The preceding outline follows the scheme developed in Jost-Yau [JY3].

As mentioned, a similar strategy works as soon as the relevant moduli space has suitable negative curvature properties. Another instance is the moduli space of $K3$ surfaces, certain complex surfaces that carry a Kähler metric with vanishing Ricci curvature according to Yau's theorem [Ya]. The moduli space for marked polarized $K3$ surfaces, or equivalently the moduli space for marked $K3$ surfaces with a Ricci-flat Kähler-Einstein metric has been constructed by Todorov [To] and Looijenga [Lj] (see also [Si2]). It is isomorphic to an open dense subset of the symmetric space of noncompact type $SO_0(3, 19)/SO(3) \times SO(19)$. Here, finiteness results were found by Hunt [Hu], but of course the general scheme just discussed also applies.

1.3 Geometric superrigidity

Let us first give the geometric version of Mostow's strong rigidity theorem:

Let G/K, G'/K' be symmetric spaces of noncompact type. Let $\Gamma \subset G$, $\Gamma' \subset G'$ be uniform lattices (i.e. discrete subgroups for which $\Gamma \backslash G$ resp. $\Gamma' \backslash G'$ is compact). After passing to subgroups of finite index, we may assume that Γ and Γ' are torsionfree, i.e. do not contain elements of finite order other than the identity, or equivalently, all elements different from the identity operate without fixed points on G/K or G'/K' resp. (Note that G operates by isometries on G/K, and so then does every subgroup of G, in particular Γ.) Therefore $\Gamma \backslash G/K$ and $\Gamma' \backslash G'/K'$ are compact manifolds, with locally symmetric metrics.

Γ is called irreducible if it does not contain a subgroup Γ_0 of finite index that splits as a nontrivial product, $\Gamma_0 = \Gamma_1 \times \Gamma_2$ (if such a splitting exists, G/K also is a nontrivial product of symmetric spaces $G_1/K_1 \times G_2/K_2$, and Γ_i operates on G_i/K_i, $i = 1, 2$.)

Theorem 1.3.1 (Mostow): *Under the above assumptions (Γ, Γ' uniform torsionfree lattices in G, G'), and if $G/K \neq Sl(2, \mathbb{R})/SO(2)$, and if Γ is irreducible, and if there exists an isomorphism*

$$\rho : \Gamma \to \Gamma'$$

then the locally symmetric spaces $\Gamma \backslash G/K$ and $\Gamma' \backslash G'/K'$ are isometric. Equivalently, ρ extends to a rational (as defined below) isomorphism

$$\rho : G \to G'.$$

Remarks:

1) The assumption is that Γ and Γ' are isomorphic as abstract groups, and the conclusion is that they are isomorphic also as subgroups of $G = G'$, i.e. conjugate subgroups of G.

2) In the appendix to § 1.1, we have normalized the metric of the symmetric space G/K. Of course, the symmetric structure is not lost if this metric is multiplied by a constant factor. In that more general situation, Mostow's theorem says that two isomorphic lattices in symmetric spaces, satisfying the assumptions stated, are isometric up to a scaling factor.

3 The condition that Γ and Γ' are torsionfree is not at all essential. In particular, ρ extends to a rational homomorphism $\rho : G \to G'$ without that condition.

4) The theorem continues to hold for nonuniform lattices (i.e. where $\Gamma \backslash G/K$ is noncompact of finite volume). This was shown in the case of rank 1 by Prasad [Pr], and in the case of rank ≥ 2 by Margulis [Mg1].

How could one go about proving such a result[1]?

Since G/K and G'/K' are simply connected complete manifolds of nonpositive curvature, by Corollary 1.1.1, all higher homotopy groups of G/K and G'/K', hence also of their quotients $\Gamma \backslash G/K$, $\Gamma \backslash G'/K'$ vanish:

$$\pi_m \left(\Gamma \backslash G/K \right) = \pi_m \left(\Gamma \backslash G'/K' \right) = \{1\} \quad \text{for all } m \geq 2.$$

Moreover, the fundamental group of $\Gamma \backslash G/K$ $\left(\Gamma \backslash G'/K' \right)$ is given by Γ (Γ'), and therefore our two spaces have isomorphic fundamental groups.

One then concludes from a not very difficult topological result that there exists a homotopy equivalence

$$f : \Gamma \backslash G/K \to \Gamma \backslash G'/K'$$

(i.e. f is continuous, and there exists a continuous $g : \Gamma \backslash G'/K' \to \Gamma \backslash G/K$ such that $f \circ g$ is homotopic to the identity of $\Gamma \backslash G'/K'$, $g \circ f$ homotopic to the identity of $\Gamma \backslash G/K$).

This is what general topology tells us. The conclusion of the theorem, however, is that there exists an isometry

$$h : \Gamma \backslash G/K \to \Gamma \backslash G'/K'.$$

We therefore have to find a method to deform a homotopy equivalence f under the above conditions into an isometry h.

It is not hard to see that it suffices to find a bijective h that is totally geodesic, i.e.

$$\nabla dh \equiv 0 \tag{1.3.1}$$

(all covariant second derivatives of h vanish, or equivalently, h maps geodesics to geodesics).

(1.3.1) is an overdetermined system of 2nd order partial differential equations for h. Therefore, given two Riemannian manifolds M, N, and a homotopy class of maps from M to N, in general, one cannot expect to find a solution of (1.3.1) within that class. (Of course, constant maps solve (1.3.1), but those are of no geometric interest.) The strategy we shall present to overcome this problem consists of two steps:

1) Solve some weaker system of 2nd order partial differential equations for h that is not overdetermined, namely the so-called harmonic map system.

---

[1]The considerations to follow will be different from Mostow's original treatment.

2) Use the special geometry of symmetric spaces to show that under the above circumstances a solution of that weaker system automatically solves the stronger system (1.3.1).

Step 1) will not use the special geometry of symmetric spaces, apart from the non-positive sectional curvature. Step 2), by contrast, will depend on certain particular identities and inequalities that only hold for symmetric spaces.

In order to include the case of lattices with torsion elements, it is convenient to generalize the setting slightly: Namely, in order to avoid difficulties with the possible singularities of the quotients $\Gamma \backslash {}^{G}\!/_{K}$, $\Gamma \backslash {}^{G'}\!/_{K'}$ in that case, we consider ρ-equivariant maps

$$f : {}^{G}\!/_{K} \to {}^{G'}\!/_{K'},$$

where "ρ-equivariant" means

$$f(\gamma x) = \rho(\gamma) f(x) \quad \forall \gamma \in \Gamma, \, x \in {}^{G}\!/_{K}.$$

When compared with the previous situation of maps between compact quotients, we are now looking at equivariant lifts of such maps to universal covers. Thus, the topological context has not changed, but we have gained a more convenient formulation.

In fact, we may consider the following general situation: Let X, Y be simply connected Riemannian manifolds with isometry groups $I(X)$ and $I(Y)$, resp. Let Γ be a discrete subgroup of $I(X)$, with fundamental region denoted by ${}^{X}\!/_{\Gamma}$, with induced volume form $d\mu_{\Gamma}$, and let

$$\rho : \Gamma \to I(Y)$$

be a homomorphism. We may then again consider ρ-equivariant maps

$$f : X \to Y,$$

i.e.

$$f(\gamma x) = \rho(\gamma) f(x) \quad \forall \gamma \in \Gamma, \, x \in X. \tag{1.3.2}$$

If Γ and $\rho(\Gamma)$ are discrete and without fixed points, i.e. $M := {}^{X}\!/_{\Gamma}$ and $N := {}^{Y}\!/_{\rho(\Gamma)}$ are manifolds with fundamental groups Γ and $\rho(\Gamma)$, resp., then any map f satisfying (1.3.2) induces a map

$$\overline{f} : M \to N,$$

and conversely any continuous map $\overline{f} : M \to N$ that induces the homomorphism ρ between $\pi_1(M) = \Gamma$ and $\pi_1(N) = \rho(\Gamma)$ lifts to a map $f : X \to Y$ satisfying (1.3.2).

For our later purposes, however, it would be inconvenient to assume that $\rho(\Gamma)$ is discrete, and so the framework of ρ-equivariance is more general.

Let us now discuss step 1) in that setting. The topology just gives us the class of (continuous) ρ-equivariant maps, whereas the geometry requires to find a map with certain "optimal" properties inside that class. Therefore, it is natural to try to find a map in that class that is as good as possible, hoping that it will turn out to be "optimal". The natural method to select particular maps in a given homotopy class would be a variational one, namely to require the map to optimize some functional. A good candidate for such a functional is the so-called energy functional which we now wish to introduce in the previously described setting. For a (sufficiently regular) ρ-equivariant map $f : X \to Y$ as above, we define

$$E(f) := \frac{1}{2} \int_{X_\Gamma} \|df(x)\|^2 d\mu_\Gamma(x) \qquad (1.3.3)$$

where $\|df(x)\|$ is the norm of the differential of f induced by the Riemannian metrics of X and Y.

In local coordinates $(x^1, \ldots, x^m)$ on X ($m = \dim X$), $(f^1, \ldots, f^n)$ on Y ($n = \dim Y$), with metrics represented by

$$\gamma_{\alpha\beta}(x) \, dx^\alpha \otimes dx^\beta$$
$$g_{ij}(f) \, df^i \otimes df^j$$

(using standard summation conventions here and in the sequel), and with $(\gamma^{\alpha\beta})$ being the inverse matrix of $(\gamma_{\alpha\beta})$, we have

$$df = \frac{\partial f^i}{\partial x^\alpha} \, dx^\alpha \otimes \frac{\partial}{\partial f^i}$$

and therefore

$$\|df(x)\|^2 = \frac{\partial f^i}{\partial x^\alpha} \frac{\partial f^j}{\partial x^\beta} \langle dx^\alpha, dx^\beta \rangle_{T^*X} \langle \frac{\partial}{\partial f^i}, \frac{\partial}{\partial f^j} \rangle_{TY}$$

$$(\langle \cdot, \cdot \rangle \text{ denoting Riemannian metrics}) \qquad (1.3.4)$$

$$= \gamma^{\alpha\beta}(x) g_{ij}(f(x)) \frac{\partial f^i}{\partial x^\alpha} \frac{\partial f^j}{\partial x^\beta}.$$

We then try to minimize $E(f)$ in the class of ρ-equivariant maps. If a smooth minimizer exists, it has to satisfy the corresponding Euler-Lagrange equations

$$\tau(f) := \text{trace} \nabla df = 0 \quad (\nabla := \text{covariant derivative in } T^*X \otimes TY), \qquad (1.3.5)$$

or, equivalently, in local coordinates

$$\frac{1}{\sqrt{\gamma}} \frac{\partial}{\partial x^\alpha} \left(\sqrt{\gamma} \gamma^{\alpha\beta} \frac{\partial f^i}{\partial x^\beta} \right) + \gamma^{\alpha\beta} \Gamma^i_{jk}(f(x)) \frac{\partial f^j}{\partial x^\alpha} \frac{\partial f^k}{\partial x^\beta} = 0 \qquad (1.3.6)$$

$$\text{for } i = 1, \ldots, n,$$

with

$$\gamma := \det(\gamma_{\alpha\beta}),$$

$$\Gamma^i_{jk} := \frac{1}{2}g^{il}\left(\frac{\partial}{\partial f^j}g_{kl} + \frac{\partial}{\partial f^k}g_{jl} - \frac{\partial}{\partial f^l}g_{jk}\right) \text{ (Christoffel symbols).}$$

Definition 1.3.1: A solution of (1.3.5) (or equivalently, of (1.3.6)) is called a *harmonic map*.

We shall later on see that the energy functional E can be minimized under the assumption that Y has nonpositive curvature, and that under that condition, minimizers are smooth, hence harmonic. Also in the situation of the Mostow rigidity theorem (and in fact under somewhat more general assumptions), harmonic maps can be shown to be totally geodesic (at present, with the exception of real hyperbolic space where a direct verification of that assertion has not yet been found so that in that case, Mostow's theorem needs a different proof), thereby also completing step 2) above.

Before embarking upon that enterprise, it will be appropriate to consider the more general statements of Margulis concerning superrigidity and arithmeticity of lattices in order to determine whether the harmonic map approach might admit corresponding generalizations. The results of Margulis will bring us into the realm of algebraic groups, and we therefore need to introduce some relevant notions. General references are the books of Zimmer [Zi] and Margulis [Mg2] himself. Here, we shall be rather brief.

Let K be an algebraically closed field of characteristic zero. Thus, in particular $\mathbb{Q} \subset K$. We let

$$Gl(n, K) := \{(n \times n) \text{ matrices } M \text{ with coefficients in } K, \det K \neq 0\}.$$

Since the determinant is a polynomial in the elements of M with rational coefficients, we say that $Gl(n, K)$ is defined over $\mathbb{Q}$. Let k be a subfield of K. An algebraic k-group is a subgroup of $Gl(n, K)$ that is the zero set of finitely many polynomials with coefficients in k. For example,

$$Sl(n, K) := \{M \in Gl(n, K) : \det M = 1\}$$

is an algebraic $\mathbb{Q}$-group. We usually say algebraic group instead of algebraic k-group.

For a ring $A \subset K$, we put

$$Gl(n, A) := \{M \in Gl(n, K) : \text{all coefficients of } M \text{ and } (\det M)^{-1} \text{ are in } A\}.$$

If G is an algebraic group, we put

$$G_A := G \cap Gl(n, A).$$

The connection between algebraic groups and semisimple Lie groups as occuring in the definition of symmetric spaces is given by

Lemma 1.3.1: *Let G be a connected, semisimple Lie group with trivial center. Then there exists a connected (semisimple) algebraic group $\widehat{G} \subset Gl(n, \mathbb{C})$ defined over $\mathbb{Q}$ such that G and $(\widehat{G}_{\mathbb{R}})^0$ (= the connected component of $\widehat{G}_{\mathbb{R}}$ containing the identity) are isomorphic as Lie groups.*

This lemma also explains the word "rational" in the statement of theorem 1.3.1. Namely, "rational" just means that when considering G and G' as algebraic groups defined over $\mathbb{Q}$, the isomorphism of the theorem is defined by equations with rational coefficients.

We shall need a little more terminology:
An algebraic k-group is called k-simple if every proper normal algebraic k-subgroup is trivial, and almost k-simple if every such subgroup is finite.
The Zariski closure of a subgroup Γ of an algebraic group G is the smallest algebraic subgroup $\overline{\Gamma}$ of G containing Γ. Γ is called Zariski dense if $\overline{\Gamma} = G$. The Borel density theorem says that in the situation of Mostow's theorem, a lattice Γ is dense in G.

If G is a semisimple algebraic group defined over k, the k-rank of G is the maximal dimension of an Abelian k-subgroup of G that can be diagonalized over k (one calls such a subgroup "k-split"). In the situation of Mostow's theorem, the rank of G/K as defined in the appendix of § 1.1 equals the $\mathbb{R}$-rank of G.

We now state the superrigidity theorem of Margulis:

Theorem 1.3.2: *Let G be a connected semisimple algebraic $\mathbb{R}$-group of $\mathbb{R}$-rank ≥ 2 for which $G_{\mathbb{R}}^0$ has no compact factors. Suppose $\Gamma \subset G_{\mathbb{R}}^0$ is an irreducible lattice. Let K be a local field[2] of characteristic 0, H a connected algebraic K-group, almost simple over K.*
Suppose

$$\rho : \Gamma \to H_K$$

is a homomorphism for which $\rho(\Gamma)$ is Zariski dense.
Then ρ factors through a homomorphism σ

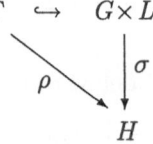

where L is a compact group.
If $K = \mathbb{R}$ and H has trivial center (i.e. H is $\mathbb{R}$-simple), ρ extends to a rational homomorphism $\rho : G \to H$.
If K is totally disconnected, then the Zariski closure of $\rho(\Gamma)$ is a compact subgroup of H.

If $K = \mathbb{Q}_p$ (which is totally disconnected), and $H = Sl(n, \mathbb{Q}_p)$, then the theorem says that $\rho(\Gamma)$ is contained in $Sl(n, \mathbb{Z}_p)$, where $\mathbb{Z}_p$ is the group of p-adic integers.

---

[2] a local field is a non-discrete, locally compact field

(Recall the definition of $\mathbb{Q}_p$: Each $r \in \mathbb{Q}$ can be uniquely written as $r = \frac{a}{b}p^m$ where $m \in \mathbb{Z}$ and a, b are prime to p. We put $\|r\|_p := p^{-m}$, and we let $\mathbb{Q}_p$ be the completion of $\mathbb{Q}$ w.r.t. $\| \cdot \|_p$, $\mathbb{Z}_p := \{s \in \mathbb{Q}_p : \|s\|_p \leq 1\}$.)

A consequence of Margulis' superrigidity theorem is his arithmeticity theorem that we now describe.

Definition 1.3.2: A lattice Γ in a connected semisimple Lie group G with trivial center and without compact factors is *arithmetic* if there exist a semisimple algebraic $\mathbb{Q}$-group H and a surjective homomorphism $\varphi : H_{\mathbb{R}}^0 \to G$ with compact kernel for which $\varphi(H_{\mathbb{Z}} \cap H_{\mathbb{R}}^0)$ and Γ are commensurable. (Γ and $\Gamma' \subset G$ are called commensurable if $\Gamma \cap \Gamma'$ is of finite index in both Γ and Γ'.)

The prototype of an arithmetic lattice is $Sl(n, \mathbb{Z}) \subset Sl(n, \mathbb{R})$, and any other arithmetic lattice can be obtained from $Sl(n, \mathbb{Z})$ in the simple manner described in Definition 1.3.2. By a theorem of Borel and Harish-Chandra, $G_{\mathbb{Z}}$ is a lattice in $G_{\mathbb{R}}$ for every semisimple algebraic group $G \subset Gl(n, \mathbb{C})$ that is defined over $\mathbb{Q}$.

Theorem 1.3.3 (Margulis): *Let G be a connected semisimple Lie group with trivial center and without compact factors and of $\mathbb{R}$-rank ≥ 2. Then any irreducible lattice in G is arithmetic.*

Remark: The arithmeticity theorem is no longer true for lattices in $H_{\mathbb{R}}^n$ and $H_{\mathbb{C}}^n$ as examples of nonarithmetic lattices show, due to Makarov [Mk], Vinberg [Vi], and Gromov-Piatetski-Shapiro [GP] for real hyperbolic spaces and to Mostow [Mo] for the complex hyperbolic space $H_{\mathbb{C}}^2$. For quaternionic hyperbolic spaces and the hyperbolic Cayley plane, however, arithmeticity continues to hold as follows from the work of Corlette [Co] and Gromov-Schoen [GS].

The question now arises whether the strategy outlined above for Mostow's theorem can be extended so as to apply to Margulis' results as well. That strategy depended crucially on the fact that the group G' in Theorem 1.3.1 was acting by isometries on a symmetric space. Furthermore, the analytic aspects of the strategy will need to exploit the nonpositive sectional curvature of that symmetric space.

Thus, the question is whether in the case of a totally disconnected field k, the group H_k can also act isometrically on some space that shares suitable geometric properties with a symmetric space of noncompact type. The answer will turn out to be positive and we wish to describe the relevant geometric construction here in a special case, namely the action of $Sl(2, K)$ on a tree where K is a totally disconnected local field, e.g. $K = \mathbb{Q}_p$.

Let thus K be a commutative field with a discrete valuation

$$v : K^* \to \mathbb{Z},$$

i.e. a homomorphism from the multiplicative group K^* onto the additive group of integers, satisfying

$$v(x + y) \geq \min(v(x), v(y)) \quad \text{for all } x, y \in K^*.$$

For $K = \mathbb{Q}_p$, the valuation is simply

$$v_p\left(\frac{a}{b}p^m\right) = m \quad \text{if } a, b \text{ are prime to } p \text{ and } m \in \mathbb{Z}.$$

For formal reasons, it is useful to put

$$v(0) := \infty.$$

We define the valuation ring

$$\mathcal{O}(v) := \{x \in K : v(x) \geq 0\}.$$

For $K = \mathbb{Q}_p$, $\mathcal{O}(v_p) = \mathbb{Z}_p$.
$\mathcal{O}(v)$ is a ring with maximal ideal generated by any $\pi \in \mathcal{O}(v)$ with $v(\pi) = 1$. For $K = \mathbb{Q}_p$, we may take $\pi = p$. The quotient $\mathcal{O}(v)/\pi\mathcal{O}(v)$ is called residue field k_v of the valuation v. For $K = \mathbb{Q}_p$, we have $k_{v_p} = \mathbb{Z}/p\mathbb{Z}$, the cyclic field with p elements. We now consider a vector space V of dimension 2 over K, and for simplicity, we identify V with K^2. Let

$$Sl(2, K) := \left\{\begin{pmatrix} a & b \\ c & d \end{pmatrix} : a, b, c, d \in K, ad - bc = 1\right\}.$$

$Sl(2, K)$ is a subgroup of $Gl(2, K)$, the group of all K-linear automorphisms of V. By a lattice of V, we mean a finitely generated $\mathcal{O}(v)$-submodule of V which generates the K-vector space V, for example $\mathcal{O}(v)^2$. A lattice therefore is a free $\mathcal{O}(v)$ module of rank 2. Two lattices L_1, L_2 are called equivalent if there exists $x \in K^*$ with $L_1 = xL_2$.

We consider the set of equivalence classes of lattices as the set of vertices of a graph, with two vertices joined by an edge if and only if the corresponding classes have representatives L_1, L_2 with the following property: There exists an $\mathcal{O}(v)$ basis (e_1, e_2) for L_1 for which $(e_1, \pi e_2)$ is an $\mathcal{O}(v)$ basis for L_2.

One verifies that in this way, a tree is obtained, i.e. a connected, nonempty graph without circuits (a circuit in a graph is a subgraph isomorphic to the graph with set of vertices $\mathbb{Z}/n\mathbb{Z}$ and edges joining i and $i+1$ for all $i \in \mathbb{Z}/n\mathbb{Z}$, for some $n \in \mathbb{N}$). We define a metric on this tree T by identifying each edge with the unit interval $[0, 1]$ with its standard metric. Thus, any two vertices are an integer distance apart and T becomes a simplicial tree. We denote the distance function on $T \times T$ by $d(\cdot, \cdot)$. Two vertices have distance n if they can be represented by lattices L and L', resp., with $L' \subset L$ and

$$L/L' \simeq \mathcal{O}(v)/\pi^n\mathcal{O}(v).$$

The vertices at distance n from the class of a lattice L_0 can be obtained as follows: $L_0/\pi^n L_0$ is a free $\mathcal{O}(v)/\pi^n\mathcal{O}(v)$ module of rank 2, and if L represents the class of a vertex at distance n from the class of L_0, then $L_0/L \simeq \mathcal{O}(v)/\pi^n\mathcal{O}(v)$ is a submodule of $L_0/\pi^n L_0$ of rank 1, i.e. a point in the projective line $\mathbb{P}(L_0/\pi^n L_0) = \mathbb{P}^1(\mathcal{O}(v)/\pi^n\mathcal{O}(v))$. In particular, for $n = 1$, the vertices of distance 1 from the vertex represented by L_0, the so-called neighbours of that vertex, correspond bijectively to the points in $\mathbb{P}(L_0/\pi L_0) = \mathbb{P}^1(\mathcal{O}(v)/\pi\mathcal{O}(v)) = \mathbb{P}^1(k_v)$. In our example $K = \mathbb{Q}_p$, $\mathbb{P}^1(k_{v_p}) = \mathbb{P}^1(\mathbb{Z}/p\mathbb{Z})$ contains $p + 1$ points. Therefore, in that case, each vertex of the corresponding tree $T = T_p$ has $p+1$ neighbors, or, equivalently, $p+1$ edges emanating from it.

(part of) the tree T_2 for $K = \mathbb{Q}_2$

$Sl(2, K)$ acts by isometries on the set of vertices of the tree T, and therefore also on the whole tree. $\begin{pmatrix} a & b \\ c & d \end{pmatrix} \in Sl(2, K)$ transforms a lattice with basis (e_1, e_2) into the lattice with basis $(ae_1 + be_2, ce_1 + de_2)$, and this action preserves equivalence classes and therefore yields an action on T, indeed. In fact, we also have an action of the larger group $Gl(2, K)$, but here elements of the form $\begin{pmatrix} a & o \\ o & a \end{pmatrix}$ act trivially; by way of contrast, all elements in $Sl(2, K)$ different from the identity act nontrivially, and therefore, we study the action of the latter group.

$Sl(2, K)$ acts transitively on the set of vertices of T, and the quotient of T by this action is an edge with its endpoints identified, i.e. the unit circle. The stabilizer of any vertex is conjugate to $Sl(2, \mathcal{O}(v))$. By the transitivity of the action, it suffices to verify this for one lattice, for example the standard lattice $\mathcal{O}(v)^2$, and for that lattice the assertion is clear.

The tree T thus constitutes a homogeneous space for the group $Sl(2, K)$, and we consider it as an analogue of the symmetric spaces encountered earlier.

Such a simplicial tree on which H_K acts is obtained whenever K is a totally disconnected local field and H_K has rank 1. In the general case of rank n, one obtains a Bruhat-Tits building, a certain simply connected metric space consisting of Euclidean simplices of dimension n glued together along boundaries, as homogeneous space for H_K. A reference is Brown, Buildings, Springer, 1989. The above discussion of trees is based on Serre, Trees, Springer, 1980.

Chapter 2

Spaces of nonpositive curvature

2.1 Local properties of Riemannian manifolds of nonpositive sectional curvature

We first recall some constructions from Riemannian geometry. A reference is J. Jost, Riemannian geometry and geometric analysis, Springer, 1995.

Let N be a complete Riemannian manifold with distance function $d(\cdot,\cdot)$, with Levi-Civita connection ∇, with $\langle\cdot,\cdot\rangle$ denoting the metric on T_xN and $\|V\|^2 = \langle V,V\rangle$ for $V \in T_xN, x \in N$, $c : [0,T] \to N$ a geodesic (parametrized proportionally to arclength, as always), $\dot{c}(t) := \frac{d}{dt}c(t)$ its tangent vector at $c(t)$. A vector field J along c (i.e. $J(t) \in T_{c(t)}N$ for all $t \in [0,T]$) is a Jacobi field if

$$\nabla_{\frac{d}{dt}}\nabla_{\frac{d}{dt}}J + R(J,\dot{c})\dot{c} = 0$$

or, with the abbreviation

$$\dot{J} := \nabla_{\frac{d}{dt}}J, \qquad \ddot{J} := \nabla_{\frac{d}{dt}}\nabla_{\frac{d}{dt}}J,$$

$$\ddot{J} + R(J,\dot{c})\dot{c} = 0. \tag{2.1.1}$$

In the sequel, a dot " $\cdot$ " will always denote a derivation w.r.t. t. The geometric interpretation of Jacobi fields is given by

Lemma 2.1.1: *Let $c : [0,T] \to N$ be geodesic, and let $c(t,s)$ be a variation of $c(t)$, i.e. $c : [0,T] \times (-\epsilon,\epsilon) \to N$ is a smooth map ($\epsilon > 0$) with $c(t,0) = c(t)$. We assume that all curves $c(\cdot,s) := c_s(\cdot)$ are geodesics.*
Then

$$J(t) = \frac{\partial}{\partial s}c(t,s)\big|_{s=0} \tag{2.1.2}$$

is a Jacobi field along $c(t) = c_0(t)$.

Conversely, every Jacobi field along $c(t)$ can be obtained by such a variation of $c(t)$ through geodesics.

The proof of the first part of Lemma 2.1.1 follows from differentiating the equation $\nabla_{\dot{c}}\dot{c} = 0$ for geodesics w.r.t. the parameter s and commuting derivatives which yields the curvature term in (2.1.1). The second part is a consequence of the existence and smooth dependence on initial data for geodesics with prescribed initial value and initial direction.

Either from this lemma, or from the fact that (2.1.1) represents a linear system of second order ODE, one deduces

Lemma 2.1.2: *Let $c : [0, T] \to N$ be geodesic. For any $V, W \in T_{c(0)}N$, there exists a unique Jacobi field J along c with*

$$J(0) = V, \qquad \dot{J}(0) = W.$$

Corollary 2.1.1: *Let $c : [0, T] \to N$ be a geodesic, $p = c(0)$, i.e.*

$$c(t) = \exp_p t\dot{c}(0),$$

where $\exp_p : T_p N \to N$ denotes the Riemannian exponential map, and let $W \in T_p N$. The Jacobi field J along c with

$$J(0) = 0, \qquad \dot{J}(0) = W$$

then is given by
$$J(t) = (D \exp_p)(t\dot{c}(0))(tW), \tag{2.1.3}$$

the derivative of $\exp_p$ evaluated at $t\dot{c}(0) \in T_p N$ and applied to tW.

The important Rauch comparison theorems control Jacobi fields via curvature bounds for N. Here, we are interested in the following special case only:

Lemma 2.1.3: *Let N be a Riemannian manifold of nonpositive sectional curvature, $c : [0, T] \to N$ a geodesic parametrized by arclength, i.e. $\|\dot{c}\| \equiv 1$. Let $J(t)$ be a Jacobi field along $c(t)$. Let*

$$f(t) := \|J(0)\| + \|\dot{J}(0)\| \cdot t, \tag{2.1.4}$$

and assume
$$f(t) > 0 \quad \text{for } 0 < t < T. \tag{2.1.5}$$

Then

$$1 \le \frac{\|J(t_1)\|}{f(t_1)} \le \frac{\|J(t_2)\|}{f(t_2)} \quad \text{for } 0 < t_1 \le t_2 < T, \tag{2.1.6}$$

and in particular

$$\|J(0)\| + \|\dot{J}(0)\| \cdot t \le \|J(t)\| \quad \text{for } 0 \le t \le T. \tag{2.1.7}$$

In Euclidean space, Jacobi fields are linear,

$$J_{eucl}(t) = J_{eucl}(0) + \dot{J}_{eucl}(0)t,$$

and the absolute value satisfies

$$|J_{eucl}(t)| = |J_{eucl}(0)| + |J_{eucl}(0)|\dot{}\,t,$$

for those Jacobi fields with $|\dot{J}_{eucl}(0)| = |J_{eucl}(0)|\dot{}$, as long as the right hand side stays nonnegative.

The geometric meaning of Lemma 2.1.3 thus is that a Jacobi field in a Riemannian manifold of nonpositive sectional curvature grows at least as fast as a Jacobi field J_{eucl} in Euclidean space with

$$\|J(0)\| = |J_{eucl}(0)|, \quad \|J(0)\|\dot{} = |J_{eucl}(0)|\dot{}$$

In particular, if $J(0) = 0$, then $\|J(t)\|$ grows at least linearly in t. From Corollary 2.1.1, we therefore obtain

Corollary 2.1.2: *Let N be a complete Riemannian manifold of nonpositive sectional curvature. Then $\exp_p$ has maximal rank everywhere on T_pN, for any $p \in N$ (note that since N is assumed to be complete, $\exp_p$ is defined on all of T_pN by the Hopf-Rinow theorem).*

In fact one has

Lemma 2.1.4: *Let N be a complete, simply connected Riemannian manifold of nonpositive sectional curvature. Then*

$$\exp_p : T_pN \to N$$

is a diffeomorphism for every $p \in N$.

This is a version of the Hadamard-Cartan Theorem 1.1.2. Instead of giving the proof here, we refer to Corollary 2.2.5 below that implies that under the conditions of Lemma 2.1.4, $\exp_p$ is bijective for every $p \in N$. Together with Corollary 2.1.2, this implies the result claimed in Lemma 2.1.4.

Lemma 2.1.5: *Let N be a complete Riemannian manifold of nonpositive sectional curvature, let $p \in N$, and suppose that*

$$\exp_p : T_pN \to N$$

is a diffeomorphism on the ball $\{V \in T_pN : \|V\| \leq \rho\}$ ($\rho > 0$). Let $\gamma : [0,1] \to N$ be a geodesic (parametrized proportionally to arclength) contained in the ball $B(p,\rho) := \{q \in N : d(p,q) \leq \rho\}$ ($= \exp_p\{\|V\| \leq \rho\}$).

Put

$$k(s) := d^2(p, \gamma(s)).$$

Then

$$k''(s) \geq 2\|\gamma'\|^2 \quad \text{for } 0 \leq s \leq 1, \quad \text{with } \gamma' := \frac{d}{ds}\gamma. \tag{2.1.8}$$

Proof: Let

$$h(x) := d^2(p, x).$$

Then

$$\mathrm{grad}h(x) = -2\exp_x^{-1}p \quad \text{for } x \in B(p, \rho), \tag{2.1.9}$$

and

$$k(s) = h(\gamma(s)).$$

Thus

$$\frac{d}{ds}k(s) = \langle(\mathrm{grad}h)(\gamma(s)), \gamma'(s)\rangle, \tag{2.1.10}$$

$$
\begin{aligned}
\frac{d^2}{ds^2}k(s) &= \langle \nabla_{\gamma'}(\mathrm{grad}h)(\gamma(s)), \gamma'(s)\rangle \tag{2.1.11}\\
&\qquad \text{since } \nabla_{\gamma'}\gamma' = 0 \text{ for a geodesic } \gamma\\
&= \nabla dh(\gamma', \gamma'),\\
&\qquad \text{by definition of the covariant Hessian } \nabla dh \text{ of } h.
\end{aligned}
$$

Let us interrupt the proof for the following

Definition 2.1.1: Let N be a Riemannian manifold, Ω open in N. $h : \Omega \to \mathbb{R}$ is called *convex* if ∇dh is positive semidefinite, and *strictly convex* if ∇dh is positive definite.

The foregoing implies that a function h is (strictly) convex iff its restriction to any geodesic is (strictly) convex.
We now continue with the proof of Lemma 2.1.5:
We consider the family of geodesics

$$c(t, s) := \exp_{\gamma(s)}\left(t\exp_{\gamma(s)}^{-1}p\right). \tag{2.1.12}$$

Thus $c(\cdot, s)$ is the geodesic from $\gamma(s) = c(0, s)$ to $p = c(1, s)$.

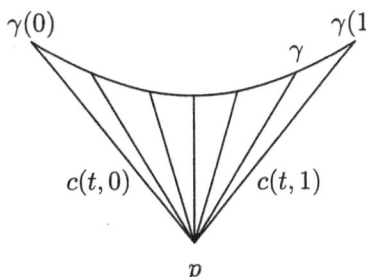

From (2.1.9), by definition of the exponential map (see Corollary 2.1.1)

$$(\mathrm{grad}h)(\gamma(s)) = -2\frac{\partial}{\partial t}c(t,s)\big|_{t=0}$$

and thus

$$\begin{aligned}\left(\nabla_{\gamma'(s)}\mathrm{grad}h\right)(\gamma(s)) &= -2\nabla_{\frac{\partial}{\partial s}}\frac{\partial}{\partial t}c(t,s)\big|_{t=0}\\ &= -2\nabla_{\frac{\partial}{\partial t}}\frac{\partial}{\partial s}c(t,s)\big|_{t=0}.\end{aligned} \qquad (2.1.13)$$

$J_s(t) := \frac{\partial}{\partial s}c(t,s)$ is a Jacobi field along the geodesic $c(\cdot,s)$ by Lemma 2.1.1, with $J_s(0) = \gamma'(s)$, $J_s(1) = 0 \in T_pN$. (2.1.13) implies

$$\left(\nabla_{\gamma'(s)}\mathrm{grad}h\right)(\gamma(s)) = -2\dot{J}_s(0).$$

This means

$$\nabla dh(\gamma'(s),\gamma'(s)) = \langle\nabla_{\gamma'(s)}\mathrm{grad}h(\gamma(s)),\gamma'(s)\rangle = -2\langle\dot{J}_s(0),J_s(0)\rangle. \qquad (2.1.14)$$

Lemma 2.1.3 gives for $t=1$

$$\|J_s(0)\| + \|J_s(0)\|^{\cdot} \leq \|J_s(1)\| = 0,$$

i.e.

$$\langle J_s(0),J_s(0)\rangle \leq -\langle\dot{J}_s(0),J_s(0)\rangle.$$

Inserting this into (2.1.14) yields

$$\nabla dh(\gamma',\gamma') \geq 2\|\gamma'\|^2,$$

and (2.1.11) then yields the claim. □

Corollary 2.1.3: *Under the assumptions of Lemma 2.1.5,*

$$\begin{aligned}d^2(p,\gamma(s)) \leq\; &(1-s)d^2(p,\gamma(0)) + sd^2(p,\gamma(1))\\ &- s(1-s)d^2(\gamma(0),\gamma(1)).\end{aligned} \qquad (2.1.15)$$

Proof: Let $k_0 : [0,1] \to \mathbb{R}$ be the function with

$$k_0(0) = d^2(p,\gamma(0)), \quad k_0(1) = d^2(p,\gamma(1)), \quad k_0''(s) = 2\|\gamma'(s)\|^2.$$

Then

$$d^2(p, \gamma(s)) \le k_0$$

as a consequence of (2.1.8). Since

$$k_0(s) = (1 - s)k_0(0) + sk_0(1) - s(1 - s)d^2(\gamma(0), \gamma(1))$$

(note $\|\gamma'(s)\| = d(\gamma(0), \gamma(1))$), the claim follows.   □

Remark: The property (2.1.15) for all $p \in N$ and all geodesics γ satisfying the assumptions of Lemma 2.1.5 is actually equivalent to the nonpositive sectional curvature of N. Namely, if the sectional curvature would be $\ge \mu > 0$ in a neighborhood of p, then locally

$$k''(s) = \nabla dh(\gamma', \gamma') \le 2\sqrt{\mu}d(p, \gamma(s))\mathrm{ctg}(\sqrt{\mu}d(p, \gamma(s))) \parallel \gamma' \parallel^2 \qquad (2.1.16)$$

(in the notation of the proof of Lemma 2.1.5) by the Rauch comparison theorem for lower curvature bounds, and (2.1.16) would yield ">" in (2.1.15) for nonconstant geodesics γ.

Lemma 2.1.5 and Corollary 2.1.3 mean that the distance function in a manifold of nonpositive curvature is at least as convex (locally) as the Euclidean distance function.

The geometric picture is

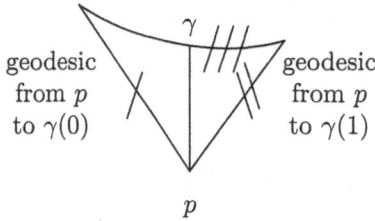

geodesic from p to $\gamma(0)$ γ geodesic from p to $\gamma(1)$

p

as compared with the Euclidean picture

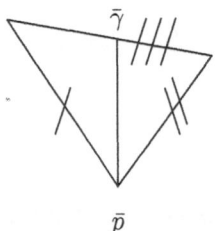

$\bar{\gamma}$

$\bar{p}$

with sides marked with the same symbol having the same length.

Let $\bar{\gamma} : [0, 1] \to \mathbb{R}^2$ be the Euclidean geodesic arc (= straight line) of the picture. Then

$$d^2(p, \gamma(s)) \le |\bar{p} - \bar{\gamma}(s)|^2 \quad \text{for } 0 \le s \le 1.$$

Corollary 2.1.4: *Let N be a complete Riemannian manifold of nonpositive sectional curvature. Let $\gamma_1 : [0,1] \to N$, $\gamma_2 : [0,1] \to N$ be geodesics with $\gamma_1(0) = \gamma_2(0) =: p$. Let $\rho > 0$ be such that $\exp_{\gamma_i(t)}$ is a diffeomorphism on the ball of radius ρ for $i = 1, 2$, $0 \le t \le 1$, and suppose*

$$d(\gamma_1(t_1), \gamma_2(t_2)) \le \rho \quad \text{for } 0 \le t_1, t_2 \le 1. \tag{2.1.17}$$

Then

$$d(\gamma_1(\tfrac{1}{2}), \gamma_2(\tfrac{1}{2})) \le \frac{1}{2} d(\gamma_1(1), \gamma_2(1)). \tag{2.1.18}$$

The geometric picture is

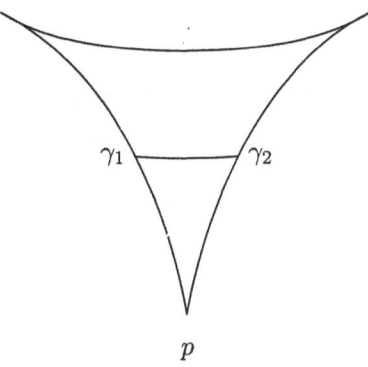

as compared with the Euclidean picture

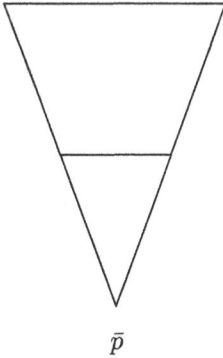

where the distance between geodesics with the same initial point grows linearly.

Proof: From (2.1.15) with $p = \gamma_1(1)$, $\gamma = \gamma_2$, $s = \frac{1}{2}$

$$d^2(\gamma_1(1), \gamma_2(\tfrac{1}{2})) \leq \frac{1}{2}d^2(\gamma_1(1), \gamma_2(1)) + \frac{1}{2}d^2(\gamma_1(1), p) \qquad (2.1.19)$$
$$- \frac{1}{4}d^2(\gamma_2(1), p).$$

From (2.1.15) with $p = \gamma_2(\tfrac{1}{2})$, $\gamma = \gamma_1$, $s = \frac{1}{2}$

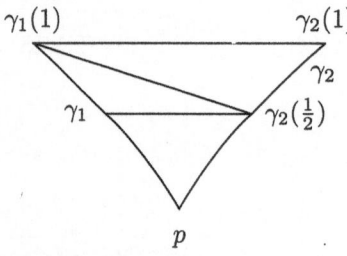

$$d^2(\gamma_1(\tfrac{1}{2}), \gamma_2(\tfrac{1}{2})) \leq \qquad (2.1.20)$$
$$\frac{1}{2}d^2(\gamma_1(1), \gamma_2(\tfrac{1}{2}))$$
$$+ \frac{1}{2}d^2(\gamma_2(\tfrac{1}{2})), p)$$
$$- \frac{1}{4}d^2(\gamma_1(1), p).$$

Inserting (2.1.20) and noting that $d(\gamma_2(\tfrac{1}{2}), p) = \frac{1}{2}d(\gamma_2(1), p)$ yields (2.1.17). □

Remarks:

1) As an exercise, you should give a direct proof of Corollary 2.1.4, using Lemma 2.1.3.

2) Again, the property of Corollary 2.1.4 is equivalent to nonpositive sectional curvature.
   Corollary 2.1.4 means that geodesics in a Riemannian manifold of nonpositive sectional curvature diverge at least as fast as Euclidean geodesics (locally).

Corollary 2.1.5: *Let N be a complete Riemannian manifold of nonpositive sectional curvature, and let γ_1, $\gamma_2 : [0,1] \to N$ be geodesics. Let $\rho > 0$ be such that $\exp_{\gamma_i(t)}$ is a diffeomorphism on the ball of radius ρ for $i = 1, 2, 0 \leq t \leq 1$, and suppose*

$$d(\gamma_1(t_1), \gamma_2(t_2)) \leq \rho \quad \text{for } 0 \leq t_1, t_2 \leq 1. \qquad (2.1.21)$$

Then

$$d(\gamma_1(t), \gamma_2(t))$$

is a convex function of $t \in [0,1]$.

Proof: We shall show

$$d(\gamma_1(\tfrac{1}{2}), \gamma_2(\tfrac{1}{2})) \leq \frac{1}{2}d(\gamma_1(0), \gamma_2(0)) + \frac{1}{2}d(\gamma_1(1), \gamma_2(1)) \qquad (2.1.22)$$

from which convexity then is easily deduced.

In order to verify (2.1.22) we let $c : [0, 1] \to N$ be the shortest geodesic from $\gamma_1(0)$ to $\gamma_2(1)$.

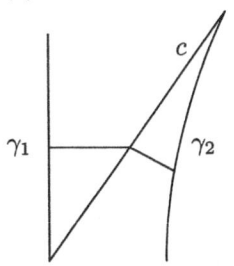

By Corollary 2.1.4, applied to the pair γ_1, c,

$$d(\gamma_1(\tfrac{1}{2}), c(\tfrac{1}{2})) \le \tfrac{1}{2} d(\gamma_1(1), \gamma_2(1)),$$
(2.1.23)

and by the same Corollary applied to c and γ_2 (with opposite orientation, i.e. $p = \gamma_2(1) = c(1)$),

$$d(c(\tfrac{1}{2}), \gamma_2(\tfrac{1}{2})) \le \tfrac{1}{2} d(\gamma_1(0), \gamma_2(0)).$$
(2.1.24)

Since $d(\gamma_1(\tfrac{1}{2}), \gamma_2(\tfrac{1}{2})) \le d(\gamma_1(\tfrac{1}{2}), c(\tfrac{1}{2})) + d(c(\tfrac{1}{2}), \gamma_2(\tfrac{1}{2}))$ by the triangle inequality, (2.1.22) follows from (2.1.23) and (2.1.24). □

Remark: Again, the statement of Corollary 2.1.5 is equivalent to nonpositive sectional curvature for a Riemannian manifold.

In the preceding, we have obtained some local characterizations of Riemannian manifolds satisfying the infinitesimal property of nonpositive sectional curvature. These characterizations are formulated solely in terms of the distance function and do not need any further properties of the Riemannian metric for their expression. Thus, they remain meaningful for more general metric spaces. Alexandrov has used the property expressed in Corollary 2.1.3 as the axiomatic basis of a theory of metric spaces of nonpositive curvature, while Busemann used the property of Corollary 2.1.4 (which is equivalent to the property of Corollary 2.1.5) (plus certain additional axioms) to build a theory of metric spaces of nonpositive curvature. In each case, a metric space is said to have nonpositive curvature, if the corresponding local property holds. We shall explore Busemann's axiom in § 2.2, and the one of Alexandrov in § 2.3. While in the Riemannian case both properties are equivalent, in general, Busemann's property is the more general one.

Before proceeding to those investigations, let us briefly mention a local property implied by strictly negative sectional curvature.

Lemma 2.1.6: *Under the assumptions and notations of Lemma 2.1.5, assume that N has sectional curvature $\le -\lambda < 0$. Then*

$$k''(s) = \nabla dh(\gamma', \gamma') \ge 2\sqrt{\lambda} d(p, \gamma(s)) \mathrm{ctgh}(\sqrt{\lambda} d(p, \gamma(s))) \, \| \, \gamma' \, \|^2 \, .$$
(2.1.25)

The proof again uses a suitable version of the Rauch comparison theorem, as the one of Lemma 2.1.5.

A consequence of Lemma 2.1.6 is

Corollary 2.1.6: *There exists some $c(\lambda) < \infty$ for $\lambda > 0$ with the following property:*

Let N be a simply connected complete Riemannian manifold of sectional cur-
vature $\leq -\lambda < 0$. Let $p \in N$, and let γ_1, $\gamma_2 : [0,1] \to N$ be geodesics with
$\gamma_1(0) = p = \gamma_2(0)$. Let $c : [0,1] \to N$ be the (unique) geodesic from $\gamma_1(1)$ to $\gamma_2(1)$.
Then for each $t \in [0,1]$, there exists $s \in [0,1]$ with

$$d(c(t), \gamma_i(s)) \leq c(\lambda) \quad \text{for } i = 1 \text{ or } 2. \tag{2.1.26}$$

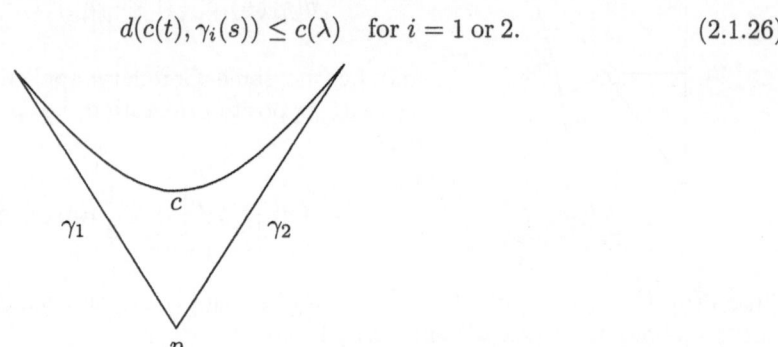

This property has been used by Gromov as an axiomatic characterization of metric
spaces of negative curvature, with important applications to group theory (see
[G3]). In this context, see also the discussion of word hyperbolic groups on p. 10.
Note, however, that for Riemannian manifolds, the property of Corollary 2.1.6 does
not imply negative, not even nonpositive sectional curvature. It should rather be
considered as a global property characterizing the behavior of negatively curved
spaces "at infinity".

Finally, simply connected complete Riemannian manifolds N of nonpositive sectional curvature can be characterized by isoperimetric inequalities. In the case of
nonpositive curvature, for $p \in N$, $n := \dim N$

$$\mathrm{Vol}_{n-1} \partial B(p, r) \geq c_n \mathrm{Vol}_n B(p, r)^{\frac{n-1}{n}} \tag{2.1.27}$$

where c_n is a positive constant depending only on n, while in the case of curvature
$\leq -\lambda < 0$, we even have

$$\mathrm{Vol}_{n-1} \partial B(p, r) \geq c(n, \lambda) \mathrm{Vol}_n B(p, r), \tag{2.1.28}$$

with $c(n, \lambda)$ depending on n and λ.
Of course, (2.1.28) is stronger than (2.1.27) only for large r.

2.2   Nonpositive curvature in the sense of Busemann

As already explained in the previous §, we now wish to build axiomatic theories of
metric spaces of nonpositive curvature by using one of the local characterizations
of nonpositive sectional curvature as a definition of that property in the context of
metric spaces. The resulting theories will be coordinate free, and they will have no

need of differentiability properties. The first such approach was found by Abraham Wald, a student of Menger, who later did fundamental work in statistics and mathematical economics. He defined curvature notions by comparing distances in geodesic quadrilaterals in a given metric space with the distances in a quadrilateral of the same side lengths in a model space of constant curvature. His approach, which worked only for surfaces, was not followed up, however, and Busemann and Alexandrov developed somewhat different notions of curvature bounds. In the present §, we consider spaces with upper curvature bound 0 in the sense of Busemann, and in the next §, we shall do the same in the context of the theory of Alexandrov.

We shall assume that all metric spaces are complete unless the contrary is explicitly stated.

Definition 2.2.1: A metric space (N, d) is called a *geodesic length space*, or simply a *geodesic space*, if for any two points x, $y \in N$, there exists a shortest geodesic arc joining them, i.e. a continuous curve $\gamma : [0,1] \to N$ with $\gamma(0) = x$, $\gamma(1) = y$, and

$$l(\gamma) = d(x, y).$$

Here, $l(\gamma)$ denotes the length of γ. It is defined as

$$l(\gamma) := \sup \left\{ \sum_{i=1}^{n} d(\gamma(t_{i-1}), \gamma(t_i)) : 0 = t_0 < t_1 < \ldots < t_n = 1, \, n \in \mathbb{N} \right\}.$$

A curve $\gamma : [0,1] \to N$ is called a geodesic if there exists $\epsilon > 0$ such that

$$l\left(\gamma_{|[t_1, t_2]}\right) = d(\gamma(t_1), \gamma(t_2)) \quad \text{whenever } |t_1 - t_2| < \epsilon.$$

This property is independent of the choice of parametrization, although the value of ϵ may change.
A geodesic $\gamma : [0,1] \to N$ is called a shortest geodesic if

$$l(\gamma) = d(\gamma(0), \gamma(1)).$$

Any curve γ of finite length, in particular any geodesic in N may be parametrized proportionally to arclength. This means, that after composition of γ with a homeomorphism $\sigma : [0,1] \to [0,1]$, the resulting curve, again called γ for simplicity, satisfies for $0 \leq t \leq 1$

$$l\left(\gamma_{|[0,t]}\right) = t l(\gamma).$$

In the sequel, we shall always assume that geodesics are parametrized proportionally to arclength.
For $x \in N$, $r > 0$, we put

$$\begin{aligned} U(x,r) &:= \{y \in N : d(x,y) < r\} \quad \text{(open ball)}, \\ B(x,r) &:= \{y \in N : d(x,y) \leq r\} \quad \text{(closed ball)}. \end{aligned}$$

A remark on terminology: A metric space is called interior, inner, or tight if the distance between any two points equals the infimum of the lengths of curves joining them. Sometimes, a space with that property has also been called length space. One should note that an interior metric space is not necessarily a geodesic length space in the sense of Definition 2.2.1, even if it is complete. A simple example of Alexander-Bishop [AB] is a sequence of intervals of length $1 + \frac{1}{n}$, with all the left endpoints identified and all the right endpoints identified. The two resulting points then are not joined by a shortest geodesic.

Definition 2.2.2: A subset S of a geodesic length space (N, d) is called *convex* if for any two points x, $y \in S$ there exists a geodesic $\gamma : [0, 1] \to N$ with $\gamma(0) = x$, $\gamma(1) = y$ and with image contained in S which is the shortest curve among all curves that connect x and y in S.
S is called *strictly convex* if for any such geodesic γ, $\gamma(t)$ is contained in the interior of S for $0 < t < 1$.

Definition 2.2.3: For x, $y \in N$, we call $m(x, y)$ a *midpoint of x and y* if

$$m(x, y) = \gamma\left(\frac{1}{2}\right)$$

for a shortest geodesic $\gamma : [0, 1] \to N$ from x to y (parametrized proportionally to arclength, as always).

Definition 2.2.4: A geodesic length space (N, d) is said to be a *Busemann nonpositive curvature (NPC) space* if for every $p \in N$ there exists $\delta_p > 0$ such that for all x, y, $z \in B(p, \delta_p)$ (and for any midpoints)

$$d(m(x, y), m(x, z)) \leq \frac{1}{2} d(y, z) \tag{2.2.1}$$

(Busemann NPC inequality).

In other words, for any two shortest geodesics γ_1, $\gamma_2 : [0, 1] \to N$ with $\gamma_1(0) = \gamma_2(0) = x \in B(p, \delta_p)$ and with endpoints $\gamma_1(1)$, $\gamma_2(1) \in B(p, \delta_p)$, we have

$$d\left(\gamma_1\left(\frac{1}{2}\right), \gamma_2\left(\frac{1}{2}\right)\right) \leq \frac{1}{2} d(\gamma_1(1), \gamma_2(1)). \tag{2.2.2}$$

Examples:

1) Let d_p be the metric on $\mathbb{R}^2$ defined by

$$d_p(x, y) := \left(|x^1 - y^1|^p + |x^2 - y^2|^p\right)^{\frac{1}{p}}$$

for $x = (x^1, x^2)$, $y = (y^1, y^2)$. For $1 < p < \infty$, the unique geodesic in $(\mathbb{R}^2, d_p)$ between any two points is given by a straight line. The homogeneity of the metric d_p implies that the Busemann NPC property holds with equality in (2.2.1).

Similarly, the Lebesgue spaces $L^p(\mathbb{R})$, $1 < p < \infty$, constitute examples of Busemann NPC spaces that are not locally compact.

Note, however, that for $p = 1$ or ∞, in the above examples, geodesics between two given points are not unique, and so the Busemann NPC inequality does not hold.

2) By Corollary 2.1.4, any complete Riemannian manifold of nonpositive sectional curvature is a Busemann NPC space.

3) Any tree, or, more generally, any Euclidean Bruhat-Tits building is a Busemann NPC space. As an exercise, the reader should verify this property for a tree.

Remark: In his book [Bu], Busemann assumed additional conditions besides (2.2.1), in particular, local compactness and the unique continuation property for geodesics. In contrast to [Bu], we shall not assume any such condition. Consequently, although we shall employ some of his reasoning, in the sequel, for certain results we needed to use different arguments, part of them being due to Gromov [G2] and Alexander-Bishop [AB].

If equality always holds in (2.2.1), one says that (N, d) is flat. If strict inequality holds in case the endpoint of neither geodesic is contained in the other one, we say that (N, d) has negative curvature (in the sense of Busemann). Note, however, that for a Riemannian manifold, negative curvature in the sense of Busemann does not imply negative sectional curvature. It only implies that the sectional curvature is nonpositive and not identically zero.

The following is immediate from the Definition 2.2.4:

Lemma 2.2.1: *Let x, $y \in (N, d)$, (N,d) a Busemann NPC space, with*

$$d(x, y) \leq \delta_x.$$

Then there is precisely one geodesic $\gamma : [0, 1] \to N$ with

$$\gamma(0) = x, \quad \gamma(1) = y, \quad d(x, y) = l(\gamma).$$

As a consequence of Lemma 2.2.1, each ball $B(x, \delta_x)$ is contractible, since if $y \in B(x, \delta_x)$, the geodesic γ of Lemma 2.2.1 is contained in $B(x, \delta_x)$ as well, and we can therefore retract $B(x, \delta_x)$ to its center x along such geodesics. In particular, it follows that a Busemann NPC space is path connected, locally path connected and locally simply connected and therefore amenable to covering space theory.

Lemma 2.2.2: *Let (N,d) be a Busemann NPC space. For $x \in N$, let*

$$\delta(x) := \sup \{\delta_x \text{ satisfying the properties of Definition 2.2.4}\}.$$

Then for x, $y \in N$

$$|\delta(x) - \delta(y)| \leq d(x, y). \tag{2.2.3}$$

In particular, if $\delta(x) = \infty$ for some $x \in N$, then $\delta(y) = \infty$ for all $y \in N$.

Proof: W.l.o.g.
$$\delta(x) \geq \delta(y).$$
(2.2.3) is obvious if $d(x,y) \geq \delta(x)$.
If $d(x,y) < \delta(x)$, then

$$B(x, \delta(x)) \supset B(y, \delta(x) - d(x,y)).$$

Therefore, the required property is satisfied in the ball $B(y, \delta(x) - d(x,y))$, hence
$\delta(y) \geq \delta(x) - d(x,y)$. □

Definition 2.2.5: A Busemann NPC space (N, d) is called *global* if $\delta(x) = \infty$ for some $x \in N$ (hence for all $x \in N$ by Lemma 2.2.2).

Theorem 2.2.1: *Let $\gamma_1\ \gamma_2 : [0,1] \to N$ be geodesics in a Busemann NPC space. If there exist shortest geodesics between $\gamma_1(t)$ and $\gamma_2(t)$ depending continuously on $0 \leq t \leq 1$, then*
$$\varphi(t) := d(\gamma_1(t), \gamma_2(t))$$
is a convex function for $0 \leq t \leq 1$. If (N,d) has negative curvature and if the two geodesics have no segment in common, $\varphi(t)$ is even strictly convex.

Proof: Denote the shortest geodesic from $\gamma_1(t)$ to $\gamma_2(t)$ by

$$c_t : [0,1] \to N \quad (c_t(0) = \gamma_1(t), \ c_t(1) = \gamma_2(t)).$$

By assumption $c_t(s)$ is continuous in t for each s, and $c_t(s)$ is of course also continuous in s. Therefore

$$S := \{c_t(s) : 0 \leq s, t \leq 1\}$$

is bounded and closed. Since $\delta(x)$ is positive and continuous in x by Lemma 2.2.2,

$$\delta := \inf\{\delta(x) : x \in S\} > 0.$$

Let
$$\mu := \max_t d(\gamma_1(t), \gamma_2(t)) = d(\gamma_1(t_0), \gamma_2(t_0)) \quad \text{for some } 0 \leq t_0 \leq 1.$$
W.l.o.g.
$$\mu > 0.$$

By continuity of $c_t(s)$, we may find $\epsilon > 0$ with

$$d(c_{t_1}(s), c_{t_2}(s)) < \frac{\delta}{2} \quad \text{whenever } |t_1 - t_2| < \epsilon, \ 0 \leq s \leq 1. \qquad (2.2.4)$$

By continuity of φ, it suffices to show

$$\varphi\left(\frac{t_1 + t_2}{2}\right) \leq \frac{1}{2}\varphi(t_1) + \frac{1}{2}\varphi(t_2) \quad \text{for } |t_1 - t_2| < \epsilon. \qquad (2.2.5)$$

For any such $t_1 + t_2$, and for $n > \frac{2\mu}{\delta}$, we put

$$p_i := c_{t_1}\left(\frac{i}{n}\right), \quad q_i := c_{t_2}\left(\frac{i}{n}\right) \quad \text{for } i = 0, 1, 2, \ldots, n.$$

In particular,

$$p_0 = \gamma_1(t_1), \quad p_n = \gamma_2(t_1),$$
$$q_0 = \gamma_1(t_2), \quad q_n = \gamma_2(t_2).$$

In order to describe the idea of the proof, let us first assume that $\gamma_1(0)$, $\gamma_1(1)$, $\gamma_2(0)$, $\gamma_2(1) \in B(p, \delta(p))$ for some $p \in N$. We may then use the argument of Corollary 2.1.6. In that case, we do not need to assume the continuous dependence of geodesics on their endpoints. Rather, in the present local situation, it follows from the statement that we are going to prove that the unique shortest geodesic depends continuously on its endpoints.

We shall verify the convexity inequality for $t = \frac{1}{2}$. It is then straightforward to deduce that inequality for $0 \le t \le 1$, first for t of the form $k2^{-n}$, $k, n \in \mathbb{N}$, and then for all t by the continuity of the distance function.

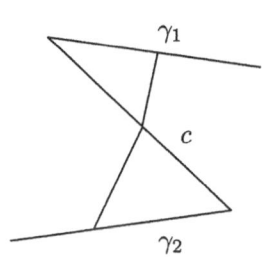

We let $c : [0, 1] \to N$ be a shortest geodesic from $\gamma_1(0)$ to $\gamma_2(1)$.

Then by the Busemann NPC inequality,

$$d(\gamma_1(\tfrac{1}{2}), c(\tfrac{1}{2})) \le \tfrac{1}{2} d(\gamma_1(1), \gamma_2(1)),$$

and also

$$d(c(\tfrac{1}{2}), \gamma_2(\tfrac{1}{2})) \le \tfrac{1}{2} d(\gamma_1(0), \gamma_2(0)).$$

Therefore, from the triangle inequality

$$d(\gamma_1(\tfrac{1}{2}), \gamma_2(\tfrac{1}{2})) \le \tfrac{1}{2} d(\gamma_1(0), \gamma_2(0)) + \tfrac{1}{2} d(\gamma_1(1), \gamma_2(1)),$$

which is the convexity inequality for $t = \frac{1}{2}$. The strict convexity assertion is also clear from the preceding construction.

We now present the actual proof.
From (2.2.4),

$$d(p_i, q_i) < \frac{\delta}{2},$$

and from the definition of μ (recalling that all geodesics c_t are parametrized proportionally to arclength) and the choice of n then

$$d(p_i, q_{i\pm1}) < \frac{\delta}{2} + \frac{\mu}{n} < \delta.$$

We may therefore apply the Busemann NPC inequality to the triples p_i, q_i, q_{i+1} and q_{i+1}, p_i, p_{i+1} to obtain

$$d(m(p_i, q_i), m(p_i, q_{i+1})) \leq \frac{1}{2}d(q_i, q_{i+1})$$

$$d(m(q_{i+1}, p_i), m(q_{i+1}, p_{i+1})) \leq \frac{1}{2}d(p_i, p_{i+1}).$$

This implies

$$d(m(p_0, q_0), m(p_n, q_n)) \leq \frac{1}{2}\left\{ \sum_{i=0}^{n-1} d(q_i, q_{i+1}) + \sum_{i=0}^{n-1} d(p_i, p_{i+1}) \right\}$$

$$= \frac{1}{2}d(q_0, q_n) + \frac{1}{2}d(p_0, p_n).$$

This is (2.2.5), since

$$m(p_0, q_0) = \gamma_1\left(\frac{t_1 + t_2}{2}\right), \quad m(p_n, q_n) = \gamma_2\left(\frac{t_1 + t_2}{2}\right).$$

The following example of Viktor Schroeder shows the necessity of the assumption that the shortest geodesic from $\gamma_1(t)$ to $\gamma_2(t)$ depends continuously on t in Theorem 2.2.1:

Let I_1, I_2 be two copies of the unit interval $[0, 1]$, with its standard metric. Join the point in I_1 with parameter value t to the point in I_2 with the same parameter value by a curve c_t of length $f(t)$, where $f : [0, 1] \to \mathbb{R}^+$ is a positive function satisfying

$$|f(t_1) - f(t_2)| < |t_1 - t_2| \quad \text{for } 0 \leq t_1 < t_2 \leq 1.$$

The resulting space then becomes a non locally compact Busemann NPC space. The curve c_t is the unique shortest geodesic connection between its endpoints, but it does not depend continuously on t. If γ_1 and γ_2 are the geodesics parametrizing I_1 and I_2 resp., then the distance function $d(\gamma_1(t), \gamma_2(t))$ is not convex unless f is a convex function.

Corollary 2.2.1: *Let (N,d) be a Busemann NPC space. Let $x \in N$, $\gamma_1, \gamma_2 : [0,1] \to N$ be geodesics with $\gamma_1(0) = x = \gamma_2(0)$. If there exist shortest geodesics between $\gamma_1(t)$ and $\gamma_2(t)$ depending continuously on $0 \le t \le 1$, then*

$$d(\gamma_1(t), \gamma_2(t)) \le t\, d(\gamma_1(1), \gamma_2(1)) \quad \text{for } 0 \le t \le 1.$$

Corollary 2.2.2: *Let (N,d) be a Busemann NPC space. Let $y \in N$, and let $\gamma : [0,1] \to N$ be a geodesic. Assume that there exist shortest geodesics from y to $\gamma(t)$ depending continuously on $0 \le t \le 1$.*
Then

$$\varphi(t) := d(\gamma(t), y)$$

is a convex function of t. If there does not exist a geodesic $\bar{\gamma} : [0,1] \to N$ containing both γ and y, then $\varphi(t)$ is even strictly convex.

Proof: The convexity follows from Theorem 2.2.1 applied to $\gamma_1(t) \equiv y$, $\gamma_2(t) = \gamma(t)$. In order to show the strict convexity, we look at the quadrilateral with corners p_1, q_1, p_0, q_0 in the proof of Theorem 2.2.1 (we shall employ the notations of the proof of Theorem 2.2.1). Since $p_0 = q_0$ in the present case, this quadrilateral degenerates into a triangle. We denote the shortest geodesic from p_1 to q_1 by $\gamma_0 : [0,1] \to N$ (thus $m(p_1, q_1) = \gamma_0(\frac{1}{2})$), and we denote the shortest geodesic from p_0 to p_1 by $c_1 : [0,1] \to N$.

Then by the NPC inequality (2.2.1)

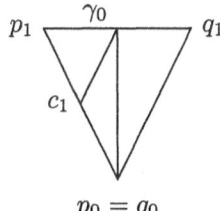

$$d(\gamma_0(\tfrac{1}{2}), c_1(\tfrac{1}{2})) \le \tfrac{1}{2} d(q_1, p_0), \qquad (2.2.6)$$

and obviously

$$d(c_1(\tfrac{1}{2}), p_0) = \tfrac{1}{2} d(p_1, p_0).$$

Therefore, from the triangle inequality

$$d(m(p_1, q_1), p_0) \le \tfrac{1}{2} d(p_1, p_0) + \tfrac{1}{2} d(q_1, p_0). \qquad (2.2.7)$$

In order to show the strict convexity, however, we need to get strict inequality in (2.2.7), hence in (2.2.6). If we had equality, then the curve consisting of the geodesic arc from $\gamma_0(\frac{1}{2})$ to $c_1(\frac{1}{2})$ and of $c_{1|[0,\frac{1}{2}]}$ would be a shortest geodesic from $\gamma_0(\frac{1}{2})$ to p_0. The same argument can be used on the other side of the triangle (denoting the shortest geodesic from p_0 to q_1 by $c_2 : [0,1] \to N$) to show that in the case of equality in (2.2.7), the curve consisting of the geodesic arc from $\gamma_0(\frac{1}{2})$ to $c_2(\frac{1}{2})$ and of $c_{2|[0,\frac{1}{2}]}$ is likewise shortest.

By local uniqueness of shortest geodesics (Lemma 2.2.1), the two curves coincide, $c_1(\frac{1}{2})$ and $c_2(\frac{1}{2})$ are both contained in the shortest geodesic c_0 from $\gamma(\frac{1}{2})$ to y. Repeating the construction with p_0' instead of p_0 and iterating, we get in the limit that p_1 or q_1 is contained in c_0. Iterating again, we obtain that p_n or q_n is contained in c_0. In this way we obtain a geodesic $\bar{\gamma}$ containing γ and y. This contradicts our assumption, and we conclude the desired strict inequality. $\square$

Remark: By a somewhat similar construction, one may also show that the function $\varphi(t)$ of Theorem 2.2.1 is strictly convex if the two geodesics γ_1, γ_2 are not contained in some other geodesic and if they are of different lengths.

Definition 2.2.6: We say that a Busemann NPC space (N,d) is *uniform* if there exists a strictly monotonic function

$$f : \mathbb{R}^+ \to \mathbb{R}^+$$

with $f(0) = 0$ and the property that whenever $y \in N$, $\gamma : [0,1] \to N$ a geodesic for which the shortest geodesic from y to $\gamma(t)$ is unique for $0 \leq t \leq 1$, we have

$$d^2(y, \gamma(\tfrac{1}{2})) \leq \frac{1}{2} d^2(y, \gamma(0)) + \frac{1}{2} d^2(y, \gamma(1)) - f(l(\gamma)). \qquad (2.2.8)$$

A Ball $B(y,r)$ in a Busemann NPC space satisfying this property for all geodesics $\gamma : [0,1] \to B(y,r)$ is called *uniformly convex*.

Corollary 2.2.3: *Let (N,d) be a Busemann NPC space. If (N,d) is locally compact, then any ball $B(y, \delta(y))$ is uniformly convex, and the convexity is also locally uniform in y (in the sense that the same function f can be chosen for all balls with center in some neighborhood of y) .*

Proof: The uniform convexity follows from Corollary 2.2.2 by a straightforward reasoning by contradiction, using local compactness to produce a limiting configuration violating Corollary 2.2.2. $\square$

From Corollary 2.2.2 and Corollary 2.2.3 it follows that a locally compact Busemann NPC space is locally uniform, and uniform if $\inf\{\delta(x) : x \in N\} > 0$. The converse, however, is far from being true, and uniformity is a much more general condition than local compactness. In what follows, we shall frequently assume uniformity, but we shall see no need for assuming local compactness.

Examples of nonuniform Busemann NPC spaces are given by the infinite Cartesian product of the spaces $(\mathbb{R}^2, d_n)$ or the spaces $(\mathbb{R}^2, d_{1+\frac{1}{n}})$, $n = 2, 3, \ldots$, as defined in Example 2.2.1.

Definition 2.2.7: For a geodesic $\gamma : [0,1] \to N$, the *capsule of radius $r > 0$ around γ* is

$$S(\gamma, r) := \{x \in N : d(x, \gamma) \leq r\},$$

with $d(x, \gamma) := \min_{0 \leq t \leq 1} d(x, \gamma(t))$.

The following result and its elegant proof are due to Alexander-Bishop [AB]:

Theorem 2.2.2: *Let (N,d) be a Busemann NPC space, $\gamma : [0,1] \to N$ a geodesic. Put*

$$\delta(\gamma) := \min_{0 \le t \le 1} \delta(\gamma(t)) \ (> 0, \text{ cf. Lemma 2.2.2}).$$

Then the capsule $S(\gamma, r)$ is convex for $0 \le r \le \frac{1}{4}\delta(\gamma)$. Furthermore, any two points in $S(\gamma, r)$ are joined by precisely one geodesic γ' that is contained in $S(\gamma, r)$ and for which $\gamma_1 = \gamma$, $\gamma_2 = \gamma'$ satisfy the assumptions of Theorem 2.2.1.

Proof: Let $\gamma_k : [0,1] \to N$ be a subsegment of γ of length $\le k\frac{r}{2}$ (w.l.o.g. $r > 0$). We shall show by induction on k that the capsule $S(\gamma_k, r)$ is convex and that for any two geodesics inside $S(\gamma_k, r)$, the assumption and conclusion of Theorem 2.2.1 hold. From the definition of $\delta(\gamma)$ it follows that this statement is correct for $k = 1$ and $k = 2$. Suppose that it is correct for some k. We shall verify that it then also holds for $k + 1$. Thus, let $\gamma_{k+1} : [0,1] \to N$ be a geodesic subsegment of γ of length $\le (k+1)\frac{r}{2}$. Let $x, y \in S(\gamma_{k+1}, r)$; we may assume $d(x, \gamma_{k+1}(0)) \le \frac{r}{2}$ and $d(y, \gamma_{k+1}(1)) \le \frac{r}{2}$, because otherwise there exists a geodesic from x to y in some $S(\gamma_k, r)$ by induction hypothesis. Put

$$p_0 := \gamma_{k+1}\left(\frac{1}{3}\right), \quad q_0 := \gamma_{k+1}\left(\frac{2}{3}\right).$$

By induction hypothesis, there exist shortest geodesics γ_{x,q_0} from x to q_0 and $\gamma_{p_0,y}$ from p_0 to y. We inductively define p_i as the midpoint of the shortest geodesic from x to q_{i-1}, and q_i as the midpoint of the shortest geodesic from p_{i-1} to y. From the NPC condition,

we have

$$d(p_{i-1}, p_i) \ \le \ \frac{r}{2^i} \quad (2.2.9)$$

$$d(q_{i-1}, q_i) \ \le \ \frac{r}{2^i},$$

and in particular $d(p_0, p_i) \le r$, $d(q_0, q_i) \le r$, for all i, so that the induction hypothesis remains applicable. These inequalities also imply that $\{p_i\}_{i \in \mathbb{N}}$ and $\{q_i\}_{i \in \mathbb{N}}$ are Cauchy sequences and hence have limit points p and q, resp., with

$$d(p, p_0) \le r, \quad d(q, q_0) \le r. \quad (2.2.10)$$

From the induction hypothesis and Corollary 2.2.1, we conclude that the geodesics γ_{x,q_i} from x to q_i converge to the geodesic $\gamma_{x,q}$ from x to q, and the geodesics $\gamma_{p_i,y}$ to $\gamma_{p,y}$. Moreover, each of these geodesics contains the geodesic from p to q. Therefore, their union gives a geodesic from x to y.

The preceding construction also implies that we may obtain geodesics $\gamma_{x,y}$ depending continuously on $x, y \in S(\gamma_{k+1}, r)$. Therefore, the assumption and conclusion of Theorem 2.2.1 inductively hold for $S(\gamma_{k+1}, r)$. Thus $S(\gamma, r)$ satisfies all the required properties. In particular, we obtain uniqueness of geodesics of the specified type inside $S(\gamma, r)$.   □

Theorem 2.2.3: *Let (N,d) be a Busemann NPC space. Let $\Gamma_0 : [0,1] \times [0,1] \to N$ be continuous with the curves $\gamma_0(t) := \Gamma_0(t,0)$ and $\gamma_1(t) := \Gamma_0(t,1)$ being geodesic. Then Γ_0 can be deformed into a geodesic homotopy, i.e. there exists a continous $\Gamma : [0,1] \times [0,1] \to N$ with*

$$\Gamma(t,0) = \gamma_0(t), \qquad \Gamma(t,1) = \gamma_1(t),$$
$$\Gamma(0,s) = \Gamma_0(0,s), \quad \Gamma(1,s) = \Gamma_0(1,s) \quad \text{for } 0 \le s, t \le 1,$$

for which the curves

$$\Gamma(\cdot, s)$$

are geodesics for all $0 \le s \le 1$.

Proof: We shall use the Alexander-Bishop argument as in the proof of Theorem 2.2.2. Let $\delta_0 := \inf\{\delta(\Gamma_0(0,s)) : 0 \le s \le 1\}$. We shall show if the result is correct if $\sup\{l(\Gamma_0(\cdot, s)) : 0 \le s \le 1\} \le k\frac{\delta_0}{4}$, then ist also holds with $\frac{3k}{2}$ in place of k. Let us first consider the case $k = 1$: We choose $\eta > 0$ so small that

$$d(\Gamma_0(t,s_1), \Gamma_0(t,s_2)) \le \frac{\delta_0}{4} \quad \text{whenever } |s_1 - s_2| \le \eta, \ 0 \le t \le 1. \qquad (2.2.11)$$

This is possible by the continuity of Γ_0.
We then have for $0 \le s_0 \le 1$, $|s - s_0| \le \eta$,

$$d(\Gamma_0(0,s_0), \Gamma_0(1,s)) \le \frac{\delta_0}{2}.$$

Therefore, the shortest geodesics $\gamma_s(t)$ from $\Gamma_0(0,s)$ to $\Gamma_0(1,s)$ are contained in $B(\Gamma(0,s_0), \frac{\delta_0}{2})$. By Lemma 2.2.1, Theorem 2.2.1 becomes applicable to conclude that

$$d(\gamma_{s_0}(t), \gamma_s(t))$$

is a convex function of t for $|s - s_0| \le \eta$.
In particular

$$d(\gamma_{s_0}(t), \gamma_s(t)) \le \frac{\delta_0}{4}$$

because this holds for $t = 0$ and $t = 1$ by (2.2.11).
Also, $\gamma_s(t)$ depends continuously on s by Theorem 2.2.2. Putting $\Gamma(s,t) := \gamma_s(t)$ therefore yields the result for $k = 1$.

Assuming the result for k, we shall now derive it for $\frac{3k}{2}$. We may again assume (2.2.11). Reparametrizing Γ_0 if necessary (so that all curves $\Gamma_0(\cdot, s)$ are parametrized proportionally to arclength), we get

$$\sup\left\{l(\Gamma_0(\cdot, s)|_{[t_1, t_2]} : 0 \le s \le 1\right\} \le k\frac{\delta_0}{4} \quad \text{whenever } 0 \le t_2 - t_1 \le \frac{2}{3}.$$

We may then apply the induction hypothesis to deform $\Gamma_0(t,s)$ for $0 \leq t \leq \frac{2}{3}$, $0 \leq s \leq 1$, into a geodesic homotopy $\Gamma_1(t,s)$, with $\Gamma_1(0,s) = \Gamma_0(0,s)$ and $\Gamma_1(\frac{2}{3},s) = \Gamma_0(\frac{2}{3},s)$ for all s. Likewise $\Gamma_1(t,s)$ for $\frac{1}{3} \leq t \leq 1$, $0 \leq s \leq 1$, may be deformed into a geodesic homotopy $\Gamma_2(t,s)$ with $\Gamma_2(\frac{1}{3},s) = \Gamma_1(\frac{1}{3},s)$, $\Gamma_2(1,s) = \Gamma_1(1,s) (= \Gamma_0(1,s))$ for all s. We then iterate this construction as in the proof of Theorem 2.2.2 to produce the desired geodesic homotopy. An important point for that construction is that we get the convexity of $d(\Gamma_m(s_1,t), \Gamma_m(s_2,t))$ for $|s_1 - s_2| \leq \eta$, $0 \leq t \leq \frac{2}{3}$ in case m is odd, $\frac{1}{3} \leq t \leq 1$ in case m is even so that (2.2.11) continuous to hold iteratively for Γ_j in place of Γ_0. □

Corollary 2.2.4: *Let (N,d) be a simply connected Busemann NPC space. Then (N,d) is a global Busemann NPC space, i.e. the NPC inequality holds for any pair of geodesics.*

Proof: Let $\gamma_1, \gamma_2 : [0,1] \to N$ be geodesics with $\gamma_1(0) = x = \gamma_2(0)$. Let $\gamma_3 : [0,1] \to N$ be a shortest geodesic from $\gamma_1(1)$ to $\gamma_2(1)$.
Since N is simply connected, we càn use Theorem 2.2.3 to construct a geodesic homotopy between γ_1 and γ_2, i.e. a continuous

$$\Gamma : [0,1] \times [0,1] \to N$$

with $\Gamma(t,0) = \gamma_1(t)$, $\Gamma(t,1) = \gamma_2(t)$, $\Gamma(0,s) = x$, $\Gamma(1,s) = \gamma_3(s)$ for $0 \leq s, t \leq 1$. Then

$$\delta := \inf\{\delta(\Gamma(t,s)) : 0 \leq s, t \leq 1\} > 0.$$

We then have from Lemma 2.2.1 and Theorem 2.2.1,

$$d(\Gamma(\tfrac{1}{2}, s_1), \Gamma(\tfrac{1}{2}, s_2)) \leq \tfrac{1}{2} d(\gamma_3(s_1), \gamma_3(s_2))$$

whenever $d(\Gamma(t,s_1), \Gamma(t,s_2)) \leq \delta$ for $0 \leq t \leq 1$. We choose $N \in \mathbb{N}$ so large that

$$d(\Gamma(t, \tfrac{i}{N}), \Gamma(t, \tfrac{i+1}{N})) \leq \delta$$

for $0 \leq t \leq 1$, $i = 0, \ldots, N-1$. Therefore

$$
\begin{aligned}
d(\gamma_1(\tfrac{1}{2}), \gamma_2(\tfrac{1}{2})) &\leq \sum_{i=0}^{N-1} d\left(\Gamma\left(\tfrac{1}{2}, \tfrac{i}{N}\right), \Gamma\left(\tfrac{1}{2}, \tfrac{i+1}{N}\right)\right) \\
&\leq \tfrac{1}{2} \sum_{i=0}^{N-1} d\left(\gamma_3\left(\tfrac{i}{N}\right), \gamma_3\left(\tfrac{i+1}{N}\right)\right) = \tfrac{1}{2} d(\gamma_3(0), \gamma_3(1)) \\
&= \tfrac{1}{2} d(\gamma_1(1), \gamma_2(1)).
\end{aligned}
$$
□

The following result ist due to Gromov [G2] and Alexander-Bishop [AB].

Corollary 2.2.5: *Let (N,d) be a simply connected Busemann NPC space. Then any x, y can be connected by precisely one geodesic arc. This geodesic arc depends continuously on x and y.*

Proof: Since $\delta(x) = \infty = \delta(y)$ by Corollary 2.2.4, the geodesic between x and y is unique by Lemma 2.2.1.

If $(x_n)_{n \in \mathbb{N}}$ converges to x, and $(y_n)_{n \in \mathbb{N}}$ to y, we may use the convexity of the distance between geodesics to show that the unique geodesic from x_n to y_n converges to the geodesic from x to y. $\square$

Corollary 2.2.6: *Let (N, d) be a simply connected Busemann NPC space, $\alpha > 1$. Then for every $y \in N$,*

$$F(x) := d^\alpha(x, y)$$

is strictly convex, i.e. for every nonconstant geodesic $\gamma : [0, 1] \to N$ and $0 < t < 1$

$$F(\gamma(t)) < (1 - t)F(\gamma(0)) + tF(\gamma(1)).$$

Proof: By Corollary 2.2.5, the assumptions of Corollary 2.2.3 are satisfied. Since $d(\gamma(t), y)$ then is strictly convex unless γ and y are contained in a common geodesic, in which case $d(\gamma(t), y)$ is linear, we easily see that $d^\alpha(\gamma(t), y)$ for $\alpha > 1$ is strictly convex if γ is nonconstant. $\square$

2.3   Nonpositive curvature in the sense of Alexandrov

We now introduce Alexandrov's notion of nonpositive curvature. It is stronger than Busemann's notion because it requires that the function $d(\gamma(t), y)$ occuring in Corollary 2.2.2 is at least as convex as the corresponding function in the Euclidean plane:

Definition 2.3.1: A geodesic length space (N, d) is said to be an *Alexandrov nonpositive curvature (NPC) space* if for every $p \in N$ there exists $\rho_p > 0$ such that for any x, y, $z \in B(p, \rho_p)$ and any shortest geodesic $\gamma : [0, 1] \to N$ with $\gamma(0) = x$, $\gamma(1) = z$, we have for $0 \le t \le 1$

$$d^2(y, \gamma(t)) \le (1 - t)d^2(y, \gamma(0)) + td^2(y, \gamma(1)) - t(1 - t)l(\gamma)^2 \qquad (2.3.1)$$

(Alexandrov NPC inequality).

Remark: In fact, it suffices to require (2.3.1) for $t = \frac{1}{2}$, analogously to Definition 2.2.4.

Note that (2.3.1) is an equality for triangles in the Euclidean plane. In fact, the geometric content of the Alexandrov NPC inequality is the following:

We construct a Euclidean triangle with vertices x', y', z' and $|x' - y'| = d(x, y)$, $|z' - y'| = d(z, y)$, $|x' - z'| = d(x, z)$, and we parametrize the edge γ' from x' to z' proportionally to arclength.

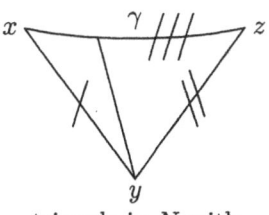
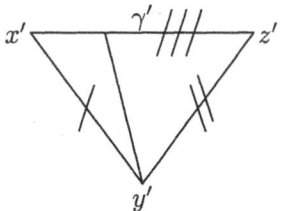

triangle in N with	triangle in the Euclidean plane
vertices x, y, z	with vertices x', y', z'

Then (2.3.1) says edges marked by the same symbol
are of the same length

$$d^2(y, \gamma(t)) \le |y' - \gamma'(t)|^2 \quad \text{for } 0 \le t \le 1.$$

Examples:

1) The spaces $(\mathbb{R}^2, d_p)$ (cf. example 1) after Def. 2.2.4) are *not* Alexandrov NPC spaces for $p \ne 2$. For $p < 2$, consider the points $x = (1, 0)$, $y = (0, 0)$, $z = (0, 1)$. Then the distance from $y = (0, 0)$ to $(\frac{1}{2}, \frac{1}{2})$ (the midpoint of x and z) is $2^{\frac{1}{p}-1} > 2^{-\frac{1}{2}}$. For $p > 2$, consider the points $x = (-1, 1)$, $y = (0, 0)$, $z = (1, 1)$. Then $d_p(x, y) = d_p(z, y) = 2^{\frac{1}{p}} < 2^{\frac{1}{2}}$ whereas the midpoint $(0, 1)$ of x and z has distance 1 from y.

   We conclude that not all Busemann NPC spaces are Alexandrov NPC spaces.

2) By Corollary 2.1.3, any complete Riemannian manifold of nonpositive sectional curvature is an Alexandrov NPC space.

3) Any Euclidean Bruhat-Tits building is an Alexandrov NPC space. The reader should verify this for a tree.

Remark: A good reference for Alexandrov's theory is Berestovskij-Nikolaev [BN]. It should be noted that the definition and comparison properties for angles play a basic rôle in Alexandrov's theory. In the approach presented here, we shall not use the concept of angles, but only consider distance comparison properties. Thus, the present § will not give full justice to Alexandrov's ideas.

As in § 2.2, we put

$$\rho(x) := \sup\{\rho_x \text{ satisfying the conditions of Definition 2.3.1}\},$$

and as in Lemma 2.2.2, we have

Lemma 2.3.1: *For an Alexandrov NPC space (N,d), and x, $y \in N$,*

$$|\rho(x) - \rho(y)| \leq d(x, y).$$

In particular, if $\rho(x) = \infty$ for some $x \in N$, then $\rho(x) = \infty$ for all $x \in N$.

The following inequalities follow from the work of Reshetnyak [Re]. Proofs of these inequalities can also be found in [KS], but we shall present a simpler proof that does not need to construct Euclidean comparison quadrilaterals.

Theorem 2.3.1: *Let (N,d) be an Alexandrov NPC space, and let γ_1, $\gamma_2 : [0,1] \to N$ be geodesics with*

$$d(\gamma_1(\tau), \gamma_2(t)) \leq \rho_{\gamma_1(\tau)}, \rho_{\gamma_2(t)} \quad \text{for } 0 \leq \tau, \, t \leq 1.$$

Then we have for $0 \leq t \leq 1$ and a parameter $0 \leq s \leq 1$.

$$
\begin{aligned}
&d^2(\gamma_1(0), \gamma_2(t)) + d^2(\gamma_1(1), \gamma_2(1-t)) \\
\leq\ & d^2(\gamma_1(0), \gamma_2(0)) + d^2(\gamma_1(1), \gamma_2(1)) + 2t^2 d^2(\gamma_2(0), \gamma_2(1)) \quad (2.3.2) \\
&+ t\left(d^2(\gamma_1(0), \gamma_1(1)) - d^2(\gamma_2(0), \gamma_2(1))\right) \\
&- ts\left(d(\gamma_1(0), \gamma_1(1)) - d(\gamma_2(0), \gamma_2(1))\right)^2 \\
&- t(1-s)\left(d(\gamma_1(0), \gamma_2(0)) - d(\gamma_1(1), \gamma_2(1))\right)^2.
\end{aligned}
$$

Note that this inequality is sharp for certain quadrilaterals in the Euclidean plane.

Proof: We first consider the case $t = 1$, $s = 0$. For simplicity of notation, we define

$$
\begin{aligned}
a_i &:= d(\gamma_i(0), \gamma_i(1)), & i &= 1, \, 2, \\
b_1 &:= d(\gamma_1(0), \gamma_2(0)), & b_2 &:= d(\gamma_1(1), \gamma_2(1)), \\
d_1 &:= d(\gamma_2(0), \gamma_1(1)), & d_2 &:= d(\gamma_1(0), \gamma_2(1)).
\end{aligned}
$$

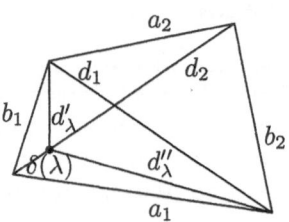

Also, we let $\delta : [0,1] \to N$ be the geodesic arc from $\gamma_1(0)$ to $\gamma_2(1)$, as always parametrized proportionally to arclength. Its length is d_2. We also put for $0 < \lambda < 1$

$$d'_\lambda := d(\gamma_2(0), \delta(\lambda)), \quad d''_\lambda := d(\gamma_1(1), \delta(\lambda)).$$

Then by the NPC inequality (2.3.1)

$$d'_\lambda{}^2 \leq (1-\lambda)b_1^2 + \lambda a_2^2 - \lambda(1-\lambda)d_2^2$$
$$d''_\lambda{}^2 \leq \lambda b_2^2 + (1-\lambda)a_1^2 - \lambda(1-\lambda)d_2^2.$$

Therefore, for $0 < \epsilon$,

$$d_1^2 \leq (d'_\lambda + d''_\lambda)^2 \leq (1+\epsilon)d'_\lambda{}^2 + \left(1 + \frac{1}{\epsilon}\right)d''_\lambda{}^2$$
$$\leq (1+\epsilon)(1-\lambda)b_1^2 + (1+\epsilon)\lambda a_2^2$$
$$+ \left(1 + \frac{1}{\epsilon}\right)\lambda b_2^2 + \left(1 + \frac{1}{\epsilon}\right)(1-\lambda)a_1^2$$
$$- \left(2 + \epsilon + \frac{1}{\epsilon}\right)\lambda(1-\lambda)d_2^2.$$

We choose $\epsilon = \frac{1-\lambda}{\lambda}$ so that the coefficient in front of d_2^2 becomes 1. This yields

$$d_2^2 + d_1^2 \leq a_1^2 + a_2^2 + \frac{1-\lambda}{\lambda}b_1^2 + \frac{\lambda}{1-\lambda}b_2^2.$$

With

$$\lambda = \frac{b_1}{b_1 + b_2},$$

we obtain

$$d_1^2 + d_2^2 \leq a_1^2 + a_2^2 + b_1^2 + b_2^2 - (b_1 - b_2)^2.$$

This is the required inequality for $t = 1$, $s = 0$. For symmetry reasons, we also obtain the inequality for $t = 1$, $s = 1$, namely

$$d_1^2 + d_2^2 \leq a_1^2 + a_2^2 + b_1^2 + b_2^2 - (a_1 - a_2)^2,$$

and taking convex combinations yields the inequality for $t = 1$, $0 \leq s \leq 1$:

$$d_1^2 + d_2^2 \leq a_1^2 + a_2^2 + b_1^2 + b_2^2 - s(a_1 - a_2)^2 - (1-s)(b_1 - b_2)^2. \qquad (2.3.3)$$

We therefore obtain the inequality for $0 \leq t \leq 1$ from the NPC inequality (2.3.1) and (2.3.3)

$$d^2(\gamma_1(0), \gamma_2(t)) + d^2(\gamma_1(1), \gamma_2(1-t))$$
$$\leq (1-t)b_1^2 + td_2^2 - t(1-t)a_2^2 + (1-t)b_1^2 + td_1^2 - t(1-t)a_2^2$$
$$\leq b_1^2 + b_2^2 + 2t^2a_2^2 - t(a_2^2 - a_1^2) - ts(a_1 - a_2)^2 - t(1-s)(b_1 - b_2)^2. \qquad \square$$

Theorem 2.3.2: *Let (N,d) be an Alexandrov NPC space, and let $\gamma_1, \gamma_2 : [0,1] \to N$ be geodesics as in Theorem 2.3.1.*
Then we have for $0 \leq t \leq 1$, $0 \leq s \leq 1$

$$d^2(\gamma_1(t), \gamma_2(t)) \leq (1-t)d^2(\gamma_1(0), \gamma_2(0)) + td^2(\gamma_1(1), \gamma_2(1)) \qquad (2.3.4)$$
$$- t(1-t)\{s(d(\gamma_1(0), \gamma_1(1)) - d(\gamma_2(0), \gamma_2(1)))^2$$
$$+ (1-s)(d(\gamma_1(0), \gamma_2(0)) - d(\gamma_1(1), \gamma_2(1)))^2\}.$$

Corollary 2.3.1: *An Alexandrov NPC space is a Busemann NPC space.*

Proof: Apply Theorem 2.3.2 to geodesics γ_1, $\gamma_2 : [0,1] \to N$ with $\gamma_1(0) = x = \gamma_2(0)$. □

Proof of Theorem 2.3.2: We shall show the inequality for $t = \frac{1}{2}$. It is then straightforward to deduce the inequality for arbitrary t.
We keep the notations of the preceding proof, and we also put

$$e_1 := d(\gamma_1(0), \gamma_2(\tfrac{1}{2})), \quad e_2 := d(\gamma_1(1), \gamma_2(\tfrac{1}{2})).$$

Then by the NPC inequality (2.3.1)

$$d^2(\gamma_1(\tfrac{1}{2}), \gamma_2(\tfrac{1}{2})) \le \frac{1}{2}e_1^2 + \frac{1}{2}e_2^2 - \frac{1}{4}a_1^2.$$

By (2.3.2)

$$e_1^2 + e_2^2 \le b_1^2 + b_2^2 + \frac{1}{2}a_1^2 - \frac{1}{2}s(a_1 - a_2)^2 - \frac{1}{2}(1-s)(b_1 - b_2)^2.$$

Thus

$$d^2(\gamma_1(\tfrac{1}{2}), \gamma_2(\tfrac{1}{2})) \le \frac{1}{2}b_1^2 + \frac{1}{2}b_2^2 - \frac{1}{4}s(a_1 - a_2)^2 - \frac{1}{4}(1-s)(b_1 - b_2)^2$$

which yields the inequality for $t = \frac{1}{2}$. □

As in § 2.2, we formulate

Definition 2.3.2: An Alexandrov NPC space (N, d) is called *global* if $\rho(x) = \infty$ for some $x \in N$ (and hence for all $x \in N$ by Lemma 2.3.1).

Corollary 2.3.2: *A simply connected Alexandrov NPC space is global, and any two points can be connected by a unique geodesic.*

Proof: Since an Alexandrov NPC space is a Busemann NPC space, the uniqueness of geodesics follows from Corollary 2.2.5, and the global property follows as in the proof of Corollary 2.2.4. □

Another useful property of Alexandrov NPC spaces is the following Pythagoras inequality

Theorem 2.3.3: *Let N be an Alexandrov NPC space, $\gamma : [0,1] \to N$ a shortest geodesic, $p \in N$ with $d(\gamma(0), p)$, $d(\gamma(1), p) < \rho(p)$.*
Suppose

$$d(\gamma(0), p) = \min_{0 \le t \le 1} d(\gamma(t), p)$$

(i.e. $\gamma(0)$ is the point on γ closest to p).
Then

$$d^2(\gamma(s), p) \ge d^2(\gamma(0), p) + s^2 d^2(\gamma(0), \gamma(1)) \quad \text{for } 0 \le s \le 1.$$

Proof: It suffices to treat the case $s = 1$.
We have from (2.3.1) that

$$d^2(\gamma(t), p) \leq (1 - t)d^2(\gamma(0), p) + td^2(\gamma(1), p) - t(1 - t)d^2(\gamma(0), \gamma(1)).$$

Since by assumption

$$d^2(\gamma(0), p) \leq d^2(\gamma(t), p),$$

we get

$$td^2(\gamma(1), p) \geq td^2(\gamma(0), p) + td^2(\gamma(0), \gamma(1)) - t^2 d^2(\gamma(0), \gamma(1)).$$

Dividing by t and letting $t \to 0$ yields the desired inequality. $\square$

Chapter 3

Convex functions and centers of mass

3.1 Minimizers of convex functions

Definition 3.1.1: Let (N, d) be a geodesic length space. A function

$$F : N \to \mathbb{R} \cup \{\infty\}$$

is called *convex* if for any geodesic arc $\gamma : [0, 1] \to N$

$$F \circ \gamma : [0, 1] \to \mathbb{R} \cup \{\infty\}$$

is convex. F is called *strictly convex* if $F \circ \gamma$ is strictly convex for any nonconstant geodesic γ.

For example, if (N, d) is a simply connected Busemann NPC space, then for every $p \in N$, $\alpha \geq 1$,

$$F(y) := d^\alpha(p, y)$$

is a convex function, by what we have shown in § 2.2, and it is strictly convex for $\alpha > 1$.

Conversely, the existence of strictly convex functions imposes restrictions on the geometry of N. For example, it excludes closed geodesics.

The following result is immediate:

Lemma 3.1.1: *Let $F : N \to \mathbb{R} \cup \{\infty\}$ be strictly convex. Then there is at most one minimizing point $p \in N$ for F, i.e. satisfying*

$$F(p) = \inf_{q \in N} F(q).$$

Proof: Let p_1, p_2 be minimizing points. Let $\gamma : [0,1] \to N$ be a geodesic with $\gamma(0) = p_1$, $\gamma(1) = p_2$. If $p_1 \neq p_2$, then by strict convexity

$$F(\gamma(\tfrac{1}{2})) < \frac{1}{2}\left(F(\gamma(0)) + F(\gamma(1))\right) = \frac{1}{2}(F(p_1) + F(p_2))$$

contradicting the minimizing property of p_1 and p_2. $\qquad\square$

A basic question for a given convex function $F : N \to \mathbb{R} \cup \{\infty\}$ is the existence of a minimizing point $p \in N$.

In order to address this question, we need to impose a convexity condition on N that is implied by nonpositive curvature:

Definition 3.1.2: We say that a geodesic length space (N, d) satisfies *condition (C)* (where "C" stands for convexity) if there exists a nonnegative lower semicontinuous function

$$\psi : N \to \mathbb{R}^+$$

with the following quantitative strict convexity condition:

For any geodesic $\gamma : [0,1] \to Y$, and any $\epsilon > 0$, there exists $\delta > 0$ such that if

$$\psi(\gamma(\tfrac{1}{2})) \geq \frac{1}{2}\psi(\gamma(0)) + \frac{1}{2}\psi(\gamma(1)) - \delta, \tag{3.1.1}$$

then

$$d(\gamma(0), \gamma(1)) < \epsilon$$

For example, if (N, d) is a uniform global Busemann NPC space, we may take $\psi(y) := d^2(p, y)$, for any $p \in N$.

Definition 3.1.3: Let $F : N \to \mathbb{R} \cup \{\infty\}$ be a function, where the geodesic length space (N, d) satisfies condition (C). For $\lambda > 0$, the *Moreau-Yosida approximation* F^λ of F is defined as

$$F^\lambda := \inf_{y \in N}(\lambda F(y) + \psi(y)).$$

Lemma 3.1.2: *Let (N,d) be complete and satisfy (C), and let $F : N \to \mathbb{R} \cup \{\infty\}$ be convex, lower semicontinuous, $\not\equiv \infty$ and bounded from below. For every $\lambda > 0$, there exists a unique $y_\lambda \in N$ with*

$$F^\lambda = \lambda F(y_\lambda) + \psi(y_\lambda). \tag{3.1.2}$$

Proof: We take a minimizing sequence $(y_n)_{n \in \mathbb{N}}$ for F^λ. This means that

$$\lim_{n \to \infty}(\lambda F(y_n) + \psi(y_n)) = F^\lambda = \inf_{y \in N}(\lambda F(y) + \psi(y)). \tag{3.1.3}$$

For $m, n \in \mathbb{N}$, we let $y_{m,n}$ be a midpoint of y_m and y_n.

The convexity of F and (C) then imply

$$
\begin{aligned}
F^\lambda \;&\le\; \lambda F(y_{m,n}) + \psi(y_{m,n}) \\
&\le\; \frac{1}{2}\lambda F(y_m) + \frac{1}{2}\lambda F(y_n) + \psi(y_{m,n}) \quad \text{by convexity} \\
&\hspace{7cm} \text{of } F
\end{aligned}
\tag{3.1.4}
$$

Condition (C),(3.1.3) and (3.1.4) imply that $(y_n)_{n\in\mathbb{N}}$ is a Cauchy sequence. Since N is complete, it converges to a unique limit y_λ. Since F and ψ are lower semicontinuous, (3.1.2) then follows from (3.1.3). $\qquad\square$

Theorem 3.1.1: *Let (N,d) be a complete geodesic length space satisfying (C), and let $F : N \to \mathbb{R} \cup \{\infty\}$ be a convex, lower semicontinuous function that is bounded from below and not identically $+\infty$. Let y_λ be as constructed in Lemma 3.1.2 for $\lambda > 0$.*
If $(\psi(y_{\lambda_n}))_{n\in\mathbb{N}}$ is bounded for some sequence $\lambda_n \to \infty$, then $(y_\lambda)_{\lambda>0}$ converges to a minimizer of F as $\lambda \to \infty$.

Proof: By definition of y_{λ_n}, y_{λ_n} minimizes $F(y) + \frac{1}{\lambda_n}\psi(y)$. Since $\psi(y_{\lambda_n})$ is bounded and λ_n tends to ∞, $(y_{\lambda_n})_{n\in\mathbb{N}}$ therefore constitutes a minimizing sequence for F. We claim that $\psi(y_\lambda)$ is a nondecreasing function of λ. To see this, let $0 < \mu < \lambda$. By definition of y_μ

$$
F(y_\lambda) + \frac{1}{\mu}\psi(y_\lambda) \ge F(y_\mu) + \frac{1}{\mu}\psi(y_\mu).
$$

This implies

$$
F(y_\lambda) + \frac{1}{\lambda}\psi(y_\lambda) \ge F(y_\mu) + \frac{1}{\lambda}\psi(y_\mu) + \left(\frac{1}{\mu} - \frac{1}{\lambda}\right) \cdot (\psi(y_\mu) - \psi(y_\lambda)).
$$

This is compatible with the definition of λ_μ only if

$$
\psi(y_\mu) \le \psi(y_\lambda)
$$

showing the claimed monotonicity property of ψ.
Since $\psi(y_\lambda)$ is bounded on the sequence $(\lambda_n)_{n\in\mathbb{N}}$ tending to ∞ and monotonic, it has to be a bounded function of $\lambda > 0$.
It follows from the definition of y_λ that

$$
F(y_\lambda) = \inf\{F(y) : \psi(y) \le \psi(y_\lambda)\}.
$$

Since $\psi(y_\lambda)$ is nondecreasing, this implies that $F(y_\lambda)$ is a nonincreasing function of λ, and as noted in the beginning, it tends to $\inf_{y\in N} F(y)$ for $\lambda \to \infty$.

Let now $\epsilon > 0$. We choose $0 < \delta$ as in (3.1.1). Since $\psi(y_\lambda)$ is monotonic and bounded in λ, we may find $\Lambda > 0$ such that for $\lambda, \mu > \Lambda$

$$
|\psi(y_\lambda) - \psi(y_\mu)| < 2\delta.
$$

If $\Lambda < \mu \leq \lambda$, we have $F(y_\mu) \geq F(y_\lambda)$ as $F(y_\lambda)$ is nonincreasing. If $y_{\mu,\lambda}$ is a midpoint of y_μ and y_λ , we obtain from the definition of y_μ

$$F(y_\mu) + \frac{1}{\mu}\psi(y_\mu) \leq F(y_{\lambda,\mu}) + \frac{1}{\mu}\psi(y_{\lambda,\mu}).$$

Also, by convexity of F, and since $F(y_\mu) \geq F(y_\lambda)$, $F(y_{\lambda,\mu}) \leq F(y_\mu)$. Therefore,

$$\psi(y_{\lambda,\mu}) \geq \psi(y_\mu) \geq \frac{1}{2}\left(\psi(y_\mu) + \psi(y_\lambda)\right) - \delta.$$

(C) then implies
$$d(y_\lambda, y_\mu) < \epsilon.$$

Thus, $(y_\lambda)_{\lambda>0}$ is a Cauchy family for $\lambda \to \infty$.
Since N is complete, there then exists a unique $y_\infty = \lim_{\lambda\to\infty} y_\lambda$. Since we have already seen that

$$\lim_{\lambda\to\infty} F(y_\lambda) = \inf_{y\in N} F(y),$$

the lower semicontinuity of F implies that

$$F(y_\infty) = \inf_{y\in N} F(y). \qquad \square$$

Obviously, the condition that $\psi(y_{\lambda_n})$ is bounded for some sequence $\lambda_n \to \infty$ is necessary as simple examples show, e.g. the convex function

$$F : \mathbb{R} \to \mathbb{R}, \quad F(x) = e^x.$$

Our assumption thus excludes that every minimizing sequence disappears at infinity in the limit. Our assumption is weaker than a coercivity assumption of the type
$$F(x_n) \to \infty \quad \text{whenever } (x_n)_{n\in\mathbb{N}} \text{ is an unbounded sequence.}$$

For example, the convex function

$$F : \mathbb{R}^2 \to \mathbb{R}, \quad F(x^1, x^2) = (x^1)^2$$

is not coercive, but still satisfies our assumption.

3.2 Centers of mass

Let (N, d) be a metric space, and let μ be a measure on N. Here and in the sequel, all measures are assumed to be nonnegative and nontrivial. We suppose that continuous functions on N are μ-measurable. Finally, we assume that

$$\mu(N) < \infty.$$

Definition 3.2.1: $q \in N$ is called a *center of mass (center of gravity, mean value)* for the measure μ if

$$\int d^2(q, x)\mu(dx) = \inf_{p \in N} \int d^2(p, x)\mu(dx) < \infty.$$

Examples:

1) If μ is a Dirac measure δ_q supported at $q \in N$, then q is its center of mass.

2) If (N, d) is a geodesic length space, and if $\mu = \delta_{q_1} + \delta_{q_2}$ is the sum of two Dirac measures, then every midpoint of q_1 and q_2 is a center of mass of μ. More generally, if $\mu = t\delta_{q_1} + (1-t)\delta_{q_2}$, $0 \le t \le 1$, and if $\gamma : [0, 1] \to N$ is a shortest geodesic with $\gamma(0) = q_1$, $\gamma(1) = q_2$, then $\gamma(1-t)$ is a center of mass of μ.
In particular, we see from this example that a center of mass need not be unique. In the present example, nonuniqueness occurs if the shortest geodesic between q_1 and q_2 is not unique.

3) If N is a Riemannian manifold, then

$$F(p) := \int d^2(p, x)\mu(dx)$$

is a differentiable function of p, with

$$DF(p) = 2 \int \exp_p^{-1}(x)\mu(dx) \quad \text{(here, } \exp_p^{-1} : N \to T_pN \text{ is considered}$$
$$\text{as vector valued function.)}$$

Thus, q is a center of mass of μ if

$$\int \exp_q^{-1}(x)\mu(dx) = 0.$$

4) Consider a tripod (N, d) obtained by taking three copies I_1, I_2, I_3 of the unit interval $[0, 1]$ and identifying $0 \in I_1$ with $0 \in I_2$ and $0 \in I_3$.

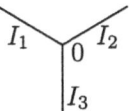 Suppose $q_i \in I_i$, $i = 1, 2, 3$, with $q_1 \le q_2 \le q_3$. If $q_2 = q_3$, then the center of mass of $\delta_{q_1} + \delta_{q_2} + \delta_{q_3}$ is 0, whereas in case $q_2 < q_3$, it is contained in the interior of I_3.

Theorem 3.2.1: Let (N, d) be a uniform, global Busemann NPC space, and let μ be a measure on N with bounded support and with $\mu(N) < \infty$.
Then there exists a unique center of mass for μ, i.e. a unique point $q \in N$ with

$$\int d^2(q, x)\mu(dx) = \inf_{p \in N} \int d^2(p, x)\mu(dx).$$

Proof: It follows from the assumptions on μ that the above infimum is finite. Since μ is nonnegative and since (N, d) is a global Busemann NPC space,

$$F(p) = \int d^2(p, x)\mu(dx)$$

is a strictly convex function on N. It is coercive in the sense that

$$F(p_i) \to \infty \quad \text{if } d^2(p_i, p_0) \to \infty \quad \text{for some fixed } p_0 \text{ and} \\ \text{a sequence } (p_i)_{i \in \mathbb{N}} \subset N,$$

since μ has bounded support.
Also, $F(p)$ is continuous.
The claim therefore follows from Theorem 3.1.1.   □

Center of mass constructions have been introduced by Elie Cartan, in order to show results of the following type:

Corollary 3.2.1: *Let (N,d) be a uniform, simply connected Busemann NPC space. Let G be a compact Lie group that acts by isometries on N, i.e. we have a map*

$$j : G \times N \quad \to \quad N \\ (g, x) \quad \to \quad gx$$

with $d(gx, gy) = d(x, y)$ for all $g \in G$, x, $y \in N$.
Then the action of G has a fixed point, i.e. there exists $q \in N$ with $gq = q$ for all $g \in G$.

Proof: Let ν be the Haar measure on G. For every $x \in N$ we then obtain a measure ν_x on the orbit $Gx = \{gx : g \in G\}$ and hence on N by putting

$$\int f(y)\nu_x(dy) = \int f(gx)\nu(dg).$$

By Theorem 3.2.1, ν_x has a unique center of mass q, i.e.

$$\int d^2(q, gx)\nu(dg) = \inf_{p \in N} \int d^2(p, gx)\nu(dg).$$

Now if $h \in G$, then

$$\int d^2(hq, gx)\nu(dg) \quad = \quad \int d^2(q, h^{-1}gx)\nu(dg) \quad \text{since } h \text{ acts isometrically}$$

$$= \quad \int d^2(q, gx)\nu(dg) \quad \text{by left invariance of the Haar measure.}$$

Therefore, since q is the unique minimizer, we must have $hq = q$ for all $h \in G$, showing that q is a fixed point for the action of G.   □

3.3 Convex hulls

Definition 3.3.1: Let (N, d) be a geodesic length space. The *convex hull* $C(A)$ of a subset A of N is the smallest convex subset of N containing A.

In general, the convex hull of a set A as defined in Definition 3.3.1 need not exist because the intersection of convex subsets of N need not be convex and so there may be more than one smallest convex set that contains A. An example is given by A being a pair of antipodal points on a sphere.

Lemma 3.3.1: *Let (N,d) be a global Busemann NPC space. Then for any $A \subset N$, the convex hull $C(A)$ exists and can be obtained as follows. Put $C_0 := A$, and for $n \in \mathbb{N}$, let C_n be the union of all geodesic arcs between points of C_{n-1}. Then*

$$C(A) = \bigcup_{n=0}^{\infty} C_n.$$

Proof: On one hand, $C(A)$ has to contain all the C_n in order to be convex and contain A. On the other hand, $C := \bigcup_{n=0}^{\infty} C_n$ is convex; namely, let $p, q \in C$. Then $p, q \in C_n$ for some n. A geodesic arc between p and q then is contained in C_{n+1}, by definition of C_{n+1}, hence also in C. Thus, C is convex. Since the geodesic arc between any two points in a uniform global Busemann NPC space is unique (see Corollary 2.2.7), C also is the smallest convex set containing A. $\square$

Lemma 3.3.2: *Let (N,d) be a global Busemann NPC space, and let $C \subset N$ be a closed, convex set, $q \in N$. Then there exists a unique $p_0 \in C$ with*

$$d(p_0, p) \le d(q, p) \text{ for all } p \in C.$$

This is called the projection of q onto C.

Proof: Since $d^2(p, \cdot)$ is a strictly convex function on the closed convex set C, it has a unique minimum $p_0 \in C$ by Theorem 3.1.1. $\square$

Lemma 3.3.3: *Let (N,d) be a global Alexandrov NPC space. Let p_0 be the projection of p onto the closed convex set C. If $p \notin C$, then for every $q \in C$*

$$d(p_0, q) < d(p, q).$$

Proof: Let $\gamma : [0, 1] \to N$ be the shortest geodesic from p_0 to q. Since C is convex, the image of γ is contained in C. Therefore, since $\gamma(0) = p_0$

$$d(\gamma(0), p) = \min_{0 \le t \le 1} d(\gamma(t), p).$$

By Theorem 2.3.3

$$d^2(q, p) \ge d^2(p_0, p) + d^2(p_0, q). \qquad \square$$

Lemma 3.3.4: *Let (N,d) be a global Alexandrov NPC space. Let μ be a measure on N with bounded support and with $\mu(N) < \infty$. Then the center of mass $q(\mu)$ of μ - which exists by Theorem 3.2.1 - is contained in the closure of the convex hull of the support of μ.*

Proof: If $q(\mu)$ were not contained in the closure C of the convex hull of the support of μ, we let q' be the projection of $q(\mu)$ onto C (note that C is convex). Then by Lemma 3.3.3

$$d^2(q',q) < d^2(q(\mu),q)$$

for every $q \in C$, hence also for every q in the support of μ and $q(\mu)$ could not be the minimum of

$$\int d^2(q,x)\mu(dx),$$

a contradiction.   □

Chapter 4

Generalized harmonic maps

4.1 The definition of generalized harmonic maps

In § 1.3, we had introduced harmonic maps between Riemannian manifolds. For a map $f : M \to N$ between Riemannian manifolds M, N, the energy was defined as

$$E(f) = \frac{1}{2} \int_M \|df(x)\|^2 d\mu(x)$$

where $d\mu$ is the measure on M induced by the Riemannian metric, df is the differential of f, and the norm $\| \cdot \|$ is induced by the Riemannian metrics of M and N. Smooth minimizers, or more generally solutions of the associated Euler-Lagrange equations, were called harmonic maps.

In view of the strategy outlined in § 1.3, we need to extend these concepts to the case where (N, d) is a geodesic length space. Obviously, if N is not a differentiable manifold, we cannot define the differential of a map f into N. Also, we do not have the norm $\| \cdot \|$ at our disposal. In order to proceed, we recall that the norm of the derivative of a smooth map may be obtained as the limit of suitable difference quotients, and we therefore attempt to define approximating energy functionals by using the distance function of N. It turns out that we can also naturally allow domains M that are more general than Riemannian manifolds.

We therefore let $(M, \mu) = (M, B, \mu)$ be a measure space, while (N, d) is a metric space. A map

$$g : M \to N$$

is called simple, if M can be covered by mutually disjoint, measurable sets $B_1, \ldots, B_k$ such that g is constant on each B_i, $i = 1, \ldots, k$. A map $f : M \to N$ is called measurable if it is the pointwise limit of a sequence of simple maps. If $f : M \to N$ is measurable and $\varphi : N \to \mathbb{R}$ is continuous, then $\varphi \circ f : M \to \mathbb{R}$ is again measurable. If f, $g : M \to N$ are measurable, then consequently $d(f(x), g(x))$ is a measurable function on M.

We let now $h : M \times M \to \mathbb{R}$ be a nonnegative, symmetric (i.e. $h(x,y) = h(y,x)$ for all x, $y \in M$) function. We then define the h-energy of $f : M \to N$ as

$$E_h(f) := \int \int h(x,y) d^2(f(x), f(y)) \mu(dy) \mu(dx).$$

With

$$e_h(f)(x) := \int h(x,y) d^2(f(x), f(y)) \mu(dy),$$

we may rewrite this as

$$E_h(f) = \int e_h(f)(x) \mu(dx).$$

In order to make contact with our previous definitions of the energy fuctional, we consider the case where M and N are Riemannian manifolds, and μ is the Riemannian measure on M. We put

$$h_\epsilon(x,y) := \chi_{U(x,\epsilon)}(y)$$

where $\chi_{U(x,\epsilon)}$ is the characteristic function of the open ball

$$U(x,\epsilon) := \{z \in M : d_M(x,z) < \epsilon\},$$

where d_M denotes the distance function of M. Thus

$$E_{h_\epsilon}(f) = \int_{x \in M} \int_{d_M(x,y) < \epsilon} d^2(f(x), f(y)) \mu(dy) \mu(dx).$$

By Taylor's theorem, there exist constants $c(\epsilon, \dim M)$ depending on ϵ and the dimension of M such that for every smooth $f : M \to N$

$$E(f) = \lim_{\epsilon \to 0} c(\epsilon, \dim M) E_{h_\epsilon}(f).$$

Since the functionals E_h can be defined in the general case where (M, μ) is a measure space and (N, d) is a metric space, we may try to define the energy functional in that case also as the limit of suitable such functionals E_h, or to work with these functionals E_h directly.

A variant of the definition that is suitable for passing to the limit $\epsilon \to 0$ would be

$$E_\epsilon(f) = \int_{x \in M} \frac{\int_{d_M(x,y) < \epsilon} d^2(f(x), f(y)) \mu(dy)}{\int_{d_M(x,y) < \epsilon} d^2(x,y) \mu(dy)} \mu(dx).$$

Since in general this cannot be written as a double integral of $d^2(f(x), f(y))$ against a symmetric kernel $h(x,y)$, minimizers of E_ϵ do not need to satisfy the mean value property of Lemma 4.1.1 anymore (this was formulated somewhat incorrectly in [Jo4]). However, one may also try

$$E'_\epsilon(f) = \int \int \frac{d^2(f(x), f(y))}{d^2(x, y)} \mu(dy)\mu(dx)$$

which has the required symmetry property, but at the expense of having a singular kernel. Thus, while E_ϵ takes finite values on suitable spaces of L^2 - mappings (as defined below), this need not be the case for E'_ϵ anymore.

Of course, in place of $\frac{1}{d^2(x,y)}$ or $\chi_{|U(x,y)}$, one may also employ a Gaussian kernel $e^{-\frac{d^2(x,y)}{\epsilon}}$. Again, in the limit, one needs to renormalize by some constant $c(\epsilon, \dim M)$, in order to obtain the energy functional for maps between Riemannian manifolds.

We first observe the following property of minimizers of E_h:

Lemma 4.1.1: $f : M \to N$ minimizes E_h iff $f(x)$ is the $h(x, \cdot)$ weighted mean value of f for a.a. $x \in M$, i.e. $f(x)$ minimizes

$$F(p) := \int h(x, y)d^2(p, f(y))\mu(dy).$$

Proof: If $f(x)$ did not minimize F, then $e_h(f)(x)$ could be decreased by replacing $f(x)$ by a minimizer q of F. Since $h(\cdot, \cdot)$ is symmetric,

$$\int e_h(f)(y)\mu(dy) \;=\; \int h(y, x)d^2(f(y), f(x))\mu(dx)\mu(dy)$$

$$=\; \int h(x, y)d^2(f(x), f(y))\mu(dy)\mu(dx)$$

could be decreased if such a replacement is possible on a set of positive measure. $\square$

It is also instructive to consider the following computation that leads to a proof of Lemma 4.1.1 in the smooth case. We consider variations

$$f_t(x) = f(x) + t\varphi(x)$$

of f. If f minimizes E_h, then

$$
\begin{aligned}
0 \;=\; \frac{d}{dt}E(f_z)\big|_{t=0} \;&=\; \frac{d}{dt}\int\int h(x, y)d^2(f_t(x), f_t(y))\mu(dy)\mu(dx)\big|_{t=0}\\
&=\; \int\int h(x, y)\, \{\nabla_1 d^2(f(x), f(y))(\varphi(x))+\\
&\qquad\qquad \nabla_2 d^2(f(x), f(y))(\varphi(y))\}\mu(dy)\mu(dx)\\
&=\; 2\int\int h(x, y)\nabla_1 d^2(f(x), f(y))\varphi(x)\mu(dy)\mu(dx)\\
&\qquad \text{because of the symmetry of } h\\
&=\; 2\int\int h(x, y)\exp^{-1}_{f(x)} f(y)\varphi(x)\mu(dy)\mu(dx).
\end{aligned}
$$

Since this has to hold for all smooth φ with compact support,

$$\int h(x,y) \exp^{-1}_{f(x)} f(y)\mu(dy) = 0$$

for all x. Of course, this just expresses that $f(x)$ is the center of mass of $f_\sharp(h(x,\cdot)\mu)$, the push forward under f of the measure $h(x,\cdot)\mu$.

The property expressed in Lemma 4.1.1 reminds us of course of the mean value property of harmonic functions on domains $\Omega \subset \mathbb{R}^m$. Namely if $f : \Omega \to \mathbb{R}$ is harmonic, then for every ball $U(x,r) \subset \Omega$

$$f(x) = \frac{1}{\mathrm{Vol}(U(x,r))} \int_{U(x,r)} f(y)dy,$$

and $f(x)$ consequently minimizes

$$F(q,r) = \int_{U(x,r)} (f(y) - q)^2 dy.$$

While this mean value property for harmonic functions holds for any radius r with $U(x,r) \subset \Omega$, in the present case, if we take for example $h_\epsilon = \chi_{|B(x,\epsilon)}$ as above, then a minimizer of E_{h_ϵ} satisfies such a mean value or center of mass property for the radius ϵ only in general.

Let us consider some other simple examples:

1) M discrete, $h(x,y) = \begin{cases} 1 & \text{if } x = y \\ 0 & \text{otherwise.} \end{cases}$

   Then for any $f : M \to N$, $E_h(f) = 0$.

2) $M = \{0, \ldots, m\} \subset \mathbb{N}$ (so that M is again discrete),

   $$h(i,j) = \begin{cases} 1 & \text{if } |i - j| \leq 1 \\ 0 & \text{otherwise.} \end{cases}$$

   Let (N,d) be a geodesic length space, $p,q \in N$. We want to minimize E_h among all maps $f : M \to N$ with $f(0) = p$, $f(m) = q$. Let $\gamma : [0,1] \to N$ be a shortest geodesic with $\gamma(0) = p$, $\gamma(1) = q$.

   A minimizing f then is given by

   $$f(i) = \gamma\left(\frac{i}{m}\right).$$

   Since consecutive points $f(i)$ and $f(i+1)$ always have the distance $\frac{L(\gamma)}{m}$, each $f(i)$ is the midpoint of $f(i-1)$ and $f(i+1)$ ($i = 1, \ldots, m-1$), and the condition of Lemma 4.1.1 is satisfied.

The same result would be obtained for

$$h(i,j) = \begin{cases} 1 & \text{for } |i - j| = 1 \\ 0 & \text{otherwise,} \end{cases}$$

because in general the value of $h(x, x)$ is irrelevant.

In order to introduce the appropriate space on which E_h is defined, it is useful to consider again the situation of ρ-equivariant maps.

We thus let (M, μ) be a measure space, and we assume that some group Γ acts on M, preserving μ (i.e. $\lambda_{\#}\mu = \mu$ for all $\lambda \in \Gamma$), and such that we get a nontrivial induced measure μ_Γ on $M/_\Gamma$. All maps in the sequel will be assumed measurable. We let (N, d) be a complete metric space with isometry group $I(N)$, and we assume that a homomorphism

$$\rho : \Gamma \to I(N)$$

is given. We then look at the class of ρ-equivariant maps

$$f : M \to N,$$

i.e. satisfying

$$f(\lambda x) = \rho(\lambda) f(x) \quad \text{for all } x \in M, \lambda \in \Gamma.$$

We can then measure the distance between two such ρ-equivariant maps f, g by putting

$$d^2(f, g) := \int d^2(f(x), g(x)) \mu_\Gamma(dx)$$

(one should think of this as integration over some kind of fundamental region for the action of Γ in M.)

If we select any ρ-equivariant map $g_0 : M \to N$ as a base map, we may put

$$L^2_\rho(M, N) := \{f : M \to N \ \rho\text{-equivariant with } d^2(f, g_0) < \infty\}.$$

Since N is a complete metric space, so then is $L^2_\rho(M, N)$.

Our generalized energies then define functionals

$$E_h : L^2_\rho(M, N) \to \mathbb{R}^+ \cup \{\infty\}.$$

The geometry of $L^2_\rho(M, N)$ can be simply described in terms of the geometry of N:

Lemma 4.1.2: *Let (N, d) be a geodesic length space. Let f_0, $f_1 \in L^2_\rho(M, N)$. For every $x \in M$, let $\gamma_x : [0, 1] \to N$ be a shortest geodesic with $\gamma_x(0) = f_0(x)$, $\gamma_x(1) = f_1(x)$, chosen equivariantly, i.e. $\rho(\lambda)\gamma_x = \gamma_{\lambda x}$ for all x, λ. Then the family of maps*

$$f_t(x) := \gamma_x(t), \quad t \in [0, 1]$$

defines a shortest geodesic in $L^2_\rho(M, N)$ between f_0 and f_1.

Proof:

$$d^2(f_0, f_t) = \int d^2(f_0(x), f_t(x))\mu_\Gamma(dx)$$

$$= \int t^2 d^2(f_0(x), f_1(x))\mu_\Gamma(dx) \quad \begin{array}{l} \text{because } \gamma_x \text{ defines} \\ \text{a shortest geodesic} \\ \text{from } f_0(x) \text{ to } f_1(x) \end{array}$$

$$= t^2 d^2(f_0, f_1).$$

This implies the shortest geodesic property.   □

Corollary 4.1.1: *Asssume (N,d) is a global (uniform) Busemann or Alexandrov NPC space. So then is $(L_\rho^2(M,N), d)$, and E_h is convex.*

Proof: The required triangle comparison properties in $(L_\rho^2(M,N), d)$ follow from those in N by integration, because of Lemma 4.1.2.
The convexity of E_h follows from the convexity of the distance between geodesics in N. Namely, in the notations of the proof of Lemma 4.1.2, one has

$$d^2(f_t(x), f_t(y)) \leq td^2(f_1(x), f_1(y)) + (1-t)d^2(f_0(x), f_0(y)). \qquad \square$$

Thus, if (N,d) is an NPC space, E_h defines a convex functional on an NPC space. Of course, $L_\rho^2(M,N)$ in general is not locally compact even if N is, but we have taken care to develop the theory of NPC spaces in chapter 2 and the existence results for minimizers of convex functionals in chapter 3 without any such local compactness assumption. Also, this has the advantage that there is no need to assume that N itself is locally compact.

Let us briefly return to the limit process $\epsilon \to 0$ addressed above and describe a general setting for such limit processes, namely the notion of Γ-limit in the sense of De Giorgi.
Let

$$F_n : Y \to \mathbb{R} \cup \{\pm\infty\}$$

be a sequence of functionals defined on a topological space Y. In order to be able to work with sequences instead of filters, we shall assume for simplicity that Y satisfies the first axiom of countability. We say that

$$F : Y \to \mathbb{R} \cup \{\pm\infty\}$$

is the Γ-limit of $(F_n)_{n\in\mathbb{N}}$,

$$F = \Gamma - \lim_{n\to\infty} F_n$$

if the following two conditions hold:

(i) whenever $(x_n)_{n\in\mathbb{N}} \subset Y$ converges to some $x \in Y$,

$$F(x) \leq \liminf_{n\to\infty} F_n(x_n);$$

(ii) for every $x \in Y$, there exists a sequence $(y_n)_{n \in \mathbb{N}} \subset Y$ converging to x and satisfying

$$F(x) \geq \limsup_{n \to \infty} F_n(y_n).$$

Γ-limits have important properties:

– If x_n is a minimizer for F_n, and if $\lim_{n \to \infty} = x$, then x is a minimizer for $F = \Gamma - \lim F_n$. Thus we may minimize a limit functional by minimizing a sequence of approximating functionals.

– Γ-limits are lower semicontinuous. As a consequence, if $F_0 : Y \to \mathbb{R} \cup \{\pm\infty\}$ is a functional that is not lower semicontinuous and if we consider the constant sequence $F_n = F_0$ for all $n \in \mathbb{N}$, then

$$\Gamma - \lim_{n \to \infty} F_n \neq F_0$$

(actually, in this case, the Γ-limit does exist, and is given by the supremum of all lower semicontinous functionals $\leq F_0$.) .

– If the F_n are all convex, then so is $F = \Gamma - \lim_{n \to \infty} F_n$.

– If Y satisfies the second axiom of countability, then every sequence $(F_n)_{n \in \mathbb{N}}$ contains a Γ-convergent subsequence. Thus, Γ-limits exist under very general circumstances.

A reference for the theory of Γ-limits is Dal Maso's book [DaM].

Thus, if we have the functionals

$$E_{h_\epsilon} : L_\rho^2(M, N) \to \mathbb{R}_+ \cup \{\infty\},$$

we put

$$\eta_\epsilon := \inf_{L_\rho^2(M,N)} E_{h_\epsilon},$$

and assuming $\eta_\epsilon \neq 0, \infty$, we may renormalize

$$\widetilde{E}_{h_\epsilon} := \frac{1}{\eta_\epsilon} E_{h_\epsilon},$$

and put

$$E := \Gamma - \lim_{\epsilon_n \to 0} \widetilde{E}_{h_{\epsilon_n}} \quad \text{for some suitable sequence } \epsilon_n \to 0 .$$

If we are dealing with (lifts to the universal covers of) mappings between compact Riemannian manifolds, one may show that E is the usual energy as defined above, up to a constant factor.

In case N is an NPC space, the E_{h_ϵ} are convex functionals, and so then is a Γ-limit E, by one of the general properties of Γ-limits stated above. Likewise, E is lower semicontinuous, because the functionals E_h are even continuous on $L_\rho^2(M, N)$.

4.2   Minimizers of generalized energy functionals

We now wish to study existence of minimizers of the functionals E_h and E introduced in the preceding §. We need to assume that E_h or E is not $\equiv \infty$. This means that we have to suppose that there exists some ρ-equivariant map f of finite energy. If $M\!\big/\!\Gamma$ is noncompact, this may be difficult to verify or even be wrong, but in many applications, $M\!\big/\!\Gamma$ is compact, and then it is usually easy.

In order to apply Theorem 3.1.1, we also need to verify that the Moreau-Yosida approximations remain bounded. This requires an additional assumption, namely that ρ is reductive, that we now wish to explain.

Definition 4.2.1: We assume that (N, d) is a Busemann NPC space. We call a subgroup G of $I(N)$ *reductive* if there exists a complete convex subset N' of N stabilized by G and with the following property:

Whenever for every countable subset G' of G, there is an unbounded sequence $(p_n)_{n \in \mathbb{N}} \subset N'$ with

$$d(p_n, gp_n) \leq const. \quad \text{for all } g \in G',$$

where the constant is independent of n, but not necessarily of g, then G stabilizes a finite-dimensional, complete, flat, totally geodesic subspace of N'. We call ρ reductive if $\rho(\Gamma)$ is.

This definition elaborates upon the ones of Jost-Yau [JY1], F. Labourie [La], and a suggestion of Scot Adams.

Let us consider some examples:

1) Let $\Sigma \subset \mathbb{R}^3$ be the surface of revolution

$$\{(x, \sin t e^{-x}, \cos t e^{-x}) : x \in \mathbb{R},\ 0 \leq t \leq 2\pi\}.$$

We consider maps $f : S^1 \to \Sigma$ in a nontrivial homotopy class, i.e. one determined by a multiple of the generator of $\pi_1(\Sigma)$. Then this homotopy class does not contain a closed geodesic. The infimum of the energy over all maps in the homotopy class is 0, but this infimum is not attained, and any minimizing sequence disappears at infinity. The corresponding representation

$$\rho : \mathbb{Z} \to I\left(\widetilde{\Sigma}\right),$$

$\widetilde{\Sigma}$ being the universal cover of Σ, a Riemannian manifold of negative curvature, is not reductive.

Similar examples can be constructed whenever the target has a "cusp".

2) Let Σ' be the surface of revolution

$$\{(x, \sin t \cosh x, \cos t \cosh x) : x \in \mathbb{R}, \ 0 \le t \le 2\pi\}.$$

Although Σ' is diffeomorphic to Σ, now every homotopy class of closed curves on Σ' contains a closed geodesic, namely a multiple of $\{(0, \sin t, \cos t) : 0 \le t \le 2\pi\}$, and thus the energy can be minimized. The corresponding representation

$$\rho : \mathbb{Z} \to I\left(\widetilde{\Sigma'}\right),$$

is reductive. Every minimizing sequence in fact is bounded.

3) Let Σ'' be the cylinder

$$\{(x, \sin t, \cos t) : x \in \mathbb{R}, \ 0 \le t \le 2\pi\}.$$

Again, closed geodesics exist; the energy can thus be minimized, and we have reductive representations of $\mathbb{Z}$ in the isometry group of the Euclidean plane, the universal cover of Σ''. Nevertheless, there do exist minimizing sequences that disappear at infinity in the limit. The Euclidean geometry of the cylinder, however, allows to bring such sequences back to compact subsets without changing their energy.

4) Let Σ''' be the surface of revolution

$$\{(x, \sin t(1 + e^{-x}), \cos t(1 + e^{-x})) : x \in \mathbb{R}, \ 0 \le t \le 2\pi\}.$$

Again, we consider maps in the same homotopy class as before. Once more, the infimum of the energy is not achieved, any minimizing sequence disappears at infinity, and the representation is not reductive.
This time, however, in contrast to 1), the infimum of the energy is positive.

5) We consider the representation

$$\rho : \mathbb{Z} \to I(\mathbb{C}),$$

where we identify $\mathbb{C}$ with the Euclidean plane that maps n to the rotation by an angle of $\frac{2\pi n}{\kappa}$ ($\kappa \in \mathbb{R}$), i.e.

$$\rho(n)z = e^{\frac{2\pi i n}{\kappa}} z.$$

The action of $\rho(\mathbb{Z})$ on $\mathbb{C}$ has a fixed point, namely 0, and therefore

$$f : \mathbb{R} \to \mathbb{C}$$

with $f(t) = 0$ for all $t \in \mathbb{R}$ is a ρ-equivariant map if $\mathbb{Z}$ acts by translations ($t \to t + n$ for $n \in \mathbb{Z}$) on $\mathbb{R}$ which has energy zero and is therefore energy minimizing.

The representation is reductive, because every sequence $(p_\nu)_{\nu \in \mathbb{N}} \subset \mathbb{C}$ with

$$\|p_\nu - e^{\frac{2\pi i n}{\kappa}} p_\nu\| \le c(n)$$

for all $n \in \mathbb{Z}$ is bounded. Also, $\rho(\mathbb{Z})$ fixes totally geodesic flat subspaces of $\mathbb{C}$, namely $\mathbb{C}$ itself and $0 \in \mathbb{C}$, the latter being a zero-dimensional such space. This example gives rise to one further remark:

If κ is rational, the quotient $\mathbb{C}/\rho(\mathbb{Z})$ is a cone, and so it has a singularity at its vertex. This cone does not have nonpositive curvature, although its cover, the Euclidean plane, has. Thus, in general, the quotient of a space of nonpositive curvature by a discrete group of isometries need not have nonpositive curvature itself.

If κ is irrational, the quotient is the half-line, and thus again has a singularity at its boundary point, but this time, it happens to have nonpositive curvature.

Remark: In the theory of algebraic groups, a group G acting on a symmetric space of noncompact type is called reductive if its Zariski closure has trivial unipotent radical, i.e. does not contain a nontrivial normal subgroup consisting of unipotent elements. One may verify that in the situation of symmetric spaces, our geometric definition of reductivity reduces to the algebraic one.

We now let N be a simply connected, uniform Busemann NPC space, and we consider a reductive representation

$$\rho : \Gamma \to I(N),$$

and we let $(f_n)_{n \in \mathbb{N}} \subset L_\rho^2(M, N)$ be a minimizing sequence for E_h, obtained as

$$f_n = J_n(f_0), \tag{4.2.1}$$

where $f_0 \in L_\rho^2(M, N)$ is assumed to satisfy

$$E_h(f_0) < \infty, \tag{4.2.2}$$

and J_n denotes the Moreau-Yosida approximation. Thus $J_n(f_0)$ is the unique minimizer of

$$E_h(f) + \frac{1}{n} d^2(f, f_0) \tag{4.2.3}$$

among $f \in L_\rho^2(M, N)$.
If

$$d^2(f_n, f_0)$$

remains bounded for $n \to \infty$, Theorem 3.1.1 yields the existence of a minimizer of E_h in $L_\rho^2(M, N)$.
In fact, the minimizer is obtained as

$$\lim_{n \to \infty} f_n.$$

Thus, let us assume

$$d^2(f_n, f_0) \to \infty \quad \text{for } n \to \infty. \tag{4.2.4}$$

We need to distinguish two cases; after choosing subsequences of $(f_n)_{n \in \mathbb{N}}$, these will constitute the only possibilities.

1) There exists $A \subset M_{/\Gamma}$ with $\mu_\Gamma(A) > 0$ such that

$$d^2(f_n(x), f_0(x)) \to \infty \quad \text{for } x \in A \tag{4.2.5}$$

uniformly, i.e. $\forall K < \infty \, \exists N < \infty \, \forall n \geq N, \, x \in A$

$$d^2(f_n(x), f_0(x)) \geq K. \tag{4.2.6}$$

We make the further assumption

$$\int_A d^2(f_n(x), \rho(\gamma) f_n(x)) \mu_\Gamma(dx) \leq c(\gamma), \tag{4.2.7}$$

where $c(\gamma)$ is a constant depending on $\gamma \in \Gamma$, but not on $n \in \mathbb{N}$, for all $\gamma \in \Gamma$. (4.2.6) and (4.2.7) imply that there exists some sequence $(x_n)_{n \in \mathbb{N}} \subset M_{/\Gamma}$ such that $(f_n(x_n))_{n \in \mathbb{N}} \subset N$ is unbounded, but that for every countable subset of $\rho(\Gamma)$,

$$d^2(f_n(x_n), \rho(\gamma) f_n(x_n))$$

is bounded independently of $n \in \mathbb{N}$ for each $\rho(\gamma)$ in that subset.

By the reductivity assumption, we conclude that $\rho(\Gamma)$ stabilizes a totally geodesic, finite dimensional, flat subspace E of N, and so we get a homomorphism

$$\rho : \Gamma \to I(E)$$

to which one may easily associate a map

$$f : M \to E$$

minimizing E_h.

We need to justify the assumption (4.2.7). (4.2.7) would follow, for example, if we had an inequality of the form

$$\int_A d^2(f_n(x), \rho(\gamma) f_n(x)) \mu_\Gamma(dx) \leq c_1(\gamma)(1 + E_h(f_n)), \tag{4.2.8}$$

again with a constant $c_1(\gamma)$ depending on γ, but not on $n \in \mathbb{N}$, for every $\gamma \in \Gamma$.

In order for (4.2.8) to hold, some kind of connectivity assumption for M is required. We call $x, y \in M$ h-equivalent

$$x \sim_h y$$

iff there exist $\delta > 0$ and open sets $U_0, \ldots, U_m$ with $\mu(U_i) > \delta$ and $x \in U_0$, $y \in U_m$, and

$$h(\xi, \eta) > 0 \quad \text{for } \xi \in U_i, \ \eta \in U_{i+1} \quad \text{for } i = 1, \ldots, m.$$

For $x \in M$, we define the h-component of x as

$$M_x := \{y \in M : y \sim_h x\}.$$

In order to make $\sim_h$ an equivalence relation, we may simply discard those $x \in M$ that are not h-equivalent to itself, because they will not cause any contribution to the energy integral. Then $\sim_h$ becomes reflexive, and it is obviously symmetric and transitive.

We also put

$$\Gamma_x := \{\gamma \in \Gamma : \gamma x \sim_h x\}.$$

If $x \sim_h y$, then also $\gamma x \sim_h \gamma y$ and consequently

$$\Gamma_x = \Gamma_y.$$

Γ_x is a subgroup of Γ operating on M_x. If M contains more than one h-component M_x of positive measure, one should consider Γ_x-equivariant maps

$$f : M_x \to N$$

and require that $\rho(\Gamma_x)$ is reductive. Namely, in that situation, the condition (4.2.8) is more natural.

Example: Consider two flat, two-dimensional tori T_1, T_2, choose points $p_i \in T_i$, and form a space X by identifying p_1 with p_2. The fundamental group Γ of X then is the free sum of the fundamental groups Γ_1 of T_1 and Γ_2 of T_2, both of course isomorphic to $\mathbb{Z}^2$. Let a_i, b_i be generators of Γ_i.
Let H^n be the n-dimensional hyperbolic space, and choose a representation

$$\rho : \Gamma \to I(H^n)$$

mapping a_i to some parabolic isometry g_i and b_i to the identity in $I(H^n)$, and assume that g_1 and g_2 are not contained in the same parabolic subgroup of $I(H^n)$. Then ρ is reductive, but $\rho(\Gamma_1)$ and $\rho(\Gamma_2)$ are not. There does not exist a ρ-equivariant harmonic map from the universal cover of X to H^n, and our minimization process will move the image of T_i towards the fixed point of g_i on the sphere at infinity of H^n. Since T_1 and T_2 are only identified at a single point, the Poincaré inequality is not valid on X, and (4.2.8) does not hold.

In this connection, it should be pointed out that the argument given in J. Jost, [Jo5], claiming that $d^2(f_n, f_n \circ \gamma) \le d^2(f_0, f_0 \circ \gamma)$ is incorrect because Lemma 4 of that paper is not applicable to $f_n \circ \gamma$.

2) For every $\epsilon > 0$, there exists $A = A_\epsilon \subset M\!\!\big/_\Gamma$ with $\mu_\Gamma(A) > \mu_\Gamma(M\!\!\big/_\Gamma) - \epsilon$ and there exists $K < \infty$ and arbitrarily large $n \in \mathbb{N}$ such that for all $x \in A$

$$d^2(f_n(x), f_0(x)) < K. \tag{4.2.9}$$

For $A \subset M\!\!\big/_\Gamma$, we put

$$E_{h,A}(f) := \inf\{E_h(g) : g = f \text{ on } A\}$$
$$d_A^2(f_1, f_2) := \int_A d^2(f_1(x), f_2(x))\mu_\Gamma(dx),$$

and we then minimize

$$E_{h,A}(f) + \frac{1}{\lambda}d_A^2(f_0, f_n)$$

for $\lambda \to \infty$. If $A = A_\epsilon$ is as above, using (4.2.9), we obtain a minimizer by Theorem 3.1.1. Furthermore, by uniqueness, we may asssume that for two such sets A_1, A_2, the corresponding minimizers coincide on $A_1 \cap A_2$. Therefore, letting $\epsilon \to 0$, we obtain a minimizer on $M\!\!\big/_\Gamma$.

Altogether, we obtain

Theorem 4.2.1: *Assume that N is a simply connected, uniform Busemann NPC space (in fact, it suffices that $d^2(p, \cdot)$ is uniformly strictly convex for every $p \in N$). Let*

$$\rho : \Gamma \to I(N)$$

be a reductive homomorphism. Assume that

$$\inf\{E_h(f) : f \in L_\rho^2(M, N)\} < \infty$$

and that (4.2.8) holds.
Then there exists a ρ-equivariant $f : M \to N$ that minimizes E_h.
The same holds for the energy E under analogous assumptions. ☐

The preceding existence theorem is based on the author's work [Jo4], [Jo5], [Jo8]. In the special case where the domain is a finite dimensional Riemannian manifold and the image is a locally compact Alexandrov NPC space, it was independently obtained by Korevaar-Schoen [KS], extending the work of Gromov-Schoen [GS] for the case of Euclidean Bruhat-Tits buildings as target. In his more recent work [Mg3] on commensurability subgroups, Margulis also obtained an existence result for generalized harmonic maps. In the Riemannian setting, i.e. where both domain and image are Riemannian manifolds, existence was shown much earlier by Eells-Sampson [ES] and Al'ber [A1], [A2], with the extension to the ρ-equivariant setting due to Diederich-Ohsawa [DO], Donaldson [Do], Corlette [Co], Jost-Yau [JY1], and

Labourie [La]. For the idea of decreasing the energy by taking averages, we should also mention Frankel's paper [Fr].

Remark: The existence theorem has been stated here for generalized harmonic maps as defined in § 4.1. This definition is based on [Jo4]. Other definitions of generalized harmonic maps have been proposed e.g. in [KS] and [Jo7]. The preceding existence proof continues to hold in those situations (again for a NPC target), because the essential point is the convexity of the energy functional.

Let us now turn to the question of uniqueness.
Let f_0, $f_1 \in L^2_\rho(M, N)$ be two minimizers of E_h. As before, we form their convex combination:
For $x \in M$, let $\gamma_x : [0,1] \to N$ be the shortest geodesic with $\gamma_x(0) = f_0(x)$, $\gamma_x(1) = f_1(x)$, and put

$$f_t(x) := \gamma_x(t). \tag{4.2.10}$$

Then, by Corollary 2.2.4 and Theorem 2.2.1

$$d^2(f_t(x), f_t(y)) \leq (1-t)d^2(f_0(x), f_0(y)) + t d^2(f_1(x), f_1(y)). \tag{4.2.11}$$

Therefore

$$
\begin{aligned}
E_h(f_t) &= \int \int h(x, y) d^2(f_t(x), f_t(y)) \mu_\Gamma(dx) \mu_\Gamma(dy) \\
&\leq (1-t) E_h(f_0) + t E_h(f_1).
\end{aligned}
\tag{4.2.12}
$$

Since $E_h(f_0)$ and $E_h(f_1)$ are minimal, we must have in fact equality. In particular, $E_h(f_t) = E_h(f_0) = E_h(f_1)$, and f_t is a minimizer as well. Actually we must also have equality in (4.2.11) for almost all x, y, hence

$$d^2(f_0(x), f_0(y)) = d^2(f_1(x), f_1(y)) = d^2(f_t(x), f_t(y)) \quad \text{for almost all } x, y.$$

Theorem 4.2.2: *Let N be a global Busemann NPC space, and let f_0, $f_1 \in L^2_\rho(M, N)$ be minimizers for E_h. If $f_t \in L^2_\rho(M, N)$ is obtained from the geodesic homotopy between f_0 and f_1, then f_t is also a minimizer for E_h. $(f_t)_{t \in [0,1]}$ constitutes a parallel family in the sense that*

$$d^2(f_t(x), f_t(y))$$

is independent of t.
In particular,

$$d^2(f_0(x), f_1(x))$$

is independent of x, again by using Theorem 2.2.1.
The same holds for E obtained as above as a Γ-limit of functionals E_{h_n} because all our convexity arguments persist.

Corollary 4.2.1: *Suppose* $\rho : \Gamma \to I(N)$ *has the property that whenever the orbits* $\rho(\Gamma)p_1$, $\rho(\Gamma)p_2$ *for* p_1, $p_2 \in N$ *are parallel in the sense that for every* $\lambda \in \Gamma$, *the geodesics*

$$\gamma_1, \gamma_2 : [0, 1] \to N$$

with $\gamma_i(0) = p_i$, $\gamma_i(1) = \rho(\lambda)p_i$ *are parallel, then* $p_1 = p_2$. *Then a minimizer for* E_h *or* E *is unique.*

The essential idea for uniqueness of course is the convexity of E, as in the previous uniqueness results of Al'ber [A2], Hartman [Ha], and Gromov-Schoen [GS].

Margulis [Mg3], observed the following consequence of uniqueness.
Let Γ_1, Γ_2 be two groups acting on M in a measure preserving manner as above, and assume that $\Gamma_1 \cap \Gamma_2$ is of finite index in both Γ_1, Γ_2. Let

$$\rho_i : \Gamma_i \to I(N)$$

be homomorphisms that coincide on $\Gamma_1 \cap \Gamma_2$, and assume that the assumption of Corollary 4.2.1 holds for the resulting

$$\rho_{1,2} : \Gamma_1 \cap \Gamma_2 \to I(N).$$

Let $f_i \in L^2_{\rho_i}(M, N)$ be minimizers for E_h. Then both f_1 and f_2 induce a minimizing g_i for E_h $(i = 1, 2)$ on $L^2_{\rho_{1,2}}(M, N)$ since $\Gamma_1 \cap \Gamma_2$ has finite index in both Γ_1, Γ_2, so that

$$E_h(g_i) = \text{index}_{\Gamma_i}(\Gamma_1 \cap \Gamma_2)E_h(f_i).$$

By uniqueness $g_1 = g_2$.
Conversely, given a minimizer for E_h in $L^2_{\rho_{1,2}}(M, N)$), it induces minimizers f_i for E_h in $L_{\rho_i}(M, N)$, $i = 1, 2$.

Using extension results for homomorphisms from lattices as described in § 1, Margulis concludes from this

Theorem 4.2.3: *Let* G *be a locally compact, compactly generated group,* Γ *a cocompact lattice in* G, Λ *a subgroup of*

$$\text{Comm}_G(\Gamma) := \{g \in G : g\Gamma^{-1}g \text{ and } \Gamma \text{ are commensurable,}$$
$$\text{meaning that their intersection has}$$
$$\text{finite index in either of them}\},$$

N a global, uniform Busemann NPC space, $\rho : \Lambda \to I(N)$ *a homomorphism satisfying the assumptions of Corollary 4.2.1.*
Then one may find a homomorphism $\tilde{\rho} : G \to I(N)$ *such that* $\tilde{\rho}_{|\Lambda} = \rho$.

Chapter 5

Bochner-Matsushima type identities for harmonic maps and rigidity theorems

5.1 The Bochner formula for harmonic one-forms and harmonic maps

In this §, we consider a harmonic map

$$f : X \to Y$$

between Riemannian manifolds. We employ the notation established in § 1.3. We wish to compute

$$\Delta \|df\|^2, \tag{5.1.1}$$

where

$$\Delta = \frac{1}{\sqrt{\gamma}} \frac{\partial}{\partial x^\alpha} \left(\sqrt{\gamma} \gamma^{\alpha\beta} \frac{\partial}{\partial x^\beta} \right) \tag{5.1.2}$$

is the Laplace-Beltrami operator of X.

We shall perform the computation in local coordinates. One may also compute in invariant notation, but this seems to be at least as complicated as the local coordinate computation (see J. Eells, L. Lemaire, [EL] for an invariant computation). In local coordinates

$$\|df(x)\|^2 = \gamma^{\alpha\beta}(x) g_{ij}(f(x)) \frac{\partial f^i}{\partial x^\alpha} \frac{\partial f^j}{\partial x^\beta}. \tag{5.1.3}$$

85

In order to simplify our computations, we choose Riemannian normal coordinates at both x and $f(x)$, in order to have for all indices

$$\gamma_{\alpha\beta} = \delta_{\alpha\beta} \quad \left(= \left\{ \begin{array}{ll} 1 & \text{for} \quad \alpha = \beta \\ 0 & \text{for} \quad \alpha \neq \beta \end{array} \right. \right) \tag{5.1.4}$$

$$\gamma_{\alpha\beta,\delta}(x) = 0 \quad \left(\gamma_{\alpha\beta,\delta} := \frac{\partial}{\partial x^\delta} \gamma_{\alpha\beta} \right) \tag{5.1.5}$$

$$g_{ij}(f(x)) = \delta_{ij} \tag{5.1.6}$$

$$g_{ij,k}(f(x)) = 0 \quad \left(g_{ij,k} := \frac{\partial}{\partial f^k} g_{ij} \right) \tag{5.1.7}$$

(In the sequel, we shall always employ the convention that an index after a comma denotes a partial derivative in the corresponding coordinate direction.)
Differentiating the relation $\gamma^{\alpha\beta}\gamma_{\beta\eta} = \delta_{\alpha\eta}$ and using (5.1.4) and (5.1.5), we get at x

$$\gamma^{\alpha\beta}_{,\delta\epsilon} = -\gamma_{\alpha\beta,\delta\epsilon} \tag{5.1.8}$$

$$\sqrt{\gamma}_{,\beta\delta} = \frac{1}{2}\gamma_{\alpha\alpha,\beta\delta}. \tag{5.1.9}$$

We recall the harmonic map equation

$$\frac{1}{\sqrt{\gamma}} \frac{\partial}{\partial x^\alpha} \left(\sqrt{\gamma} \gamma^{\alpha\beta} f^i_{x^\beta} \right) + \gamma^{\alpha\beta} \Gamma^i_{jk}(f(x)) f^j_{x^\alpha} f^k_{x^\beta} = 0, \tag{5.1.10}$$

abbreviating $f^i_{x^\beta} = \frac{\partial f^i}{\partial x^\beta}$.

We differentiate (5.1.10) at x w.r.t. x^δ and obtain, using (5.1.4)-(5.1.9)

$$f^i_{x^\alpha x^\alpha x^\delta} - \gamma_{\alpha\beta,\alpha\delta} f^i_{x^\beta} + \frac{1}{2}\gamma_{\alpha\alpha,\beta\delta} f^i_{x^\beta} + \frac{1}{2}\left(g_{ij,kl} + g_{ik,jl} - g_{jk,il} \right) f^l_{x^\delta} f^j_{x^\alpha} f^k_{x^\beta} = 0. \tag{5.1.11}$$

From (5.1.8), we have

$$\Delta\gamma^{\alpha\beta} = -\gamma_{\alpha\beta,\delta\delta} \tag{5.1.12}$$

and from the chain rule

$$\Delta g_{ij}(f(x)) = g_{ij,kl} f^k_{x^\delta} f^l_{x^\delta}. \tag{5.1.13}$$

From (5.1.11)–(5.1.13), we obtain

$$\Delta \left(\gamma^{\alpha\beta}(x) g_{ij}(f(x)) f^i_{x^\alpha} f^j_{x^\beta} \right) = \tag{5.1.14}$$

$$2 f^i_{x^\alpha x^\beta} f^i_{x^\alpha x^\delta} - \left(\gamma_{\delta\delta,\beta\alpha} + \gamma_{\alpha\beta,\delta\delta} - \gamma_{\delta\alpha,\delta\beta} - \gamma_{\delta\beta,\delta\alpha} \right) f^i_{x^\alpha} f^i_{x^\beta}$$

$$+ \left(g_{ij,kl} + g_{lk,ij} - g_{ik,lj} - g_{jl,ki} \right) f^i_{x^\alpha} f^j_{x^\alpha} f^k_{x^\delta} f^l_{x^\delta},$$

where we have interchanged the indices α and δ in (5.1.11) (and consequently also l and j), and where we have used symmetries like

$$\gamma_{\delta\alpha,\delta\beta}f^i_{x^\alpha}f^i_{x^\beta} = \gamma_{\delta\beta,\delta\alpha}f^i_{x^\alpha}f^i_{x^\beta}, \qquad g_{il,kj}f^i_{x^\alpha}f^j_{x^\alpha} = g_{jl,ki}f^i_{x^\alpha}f^j_{x^\alpha},$$

and of course also

$$\gamma_{\alpha\beta} = \gamma_{\beta\alpha}, \quad \gamma_{\alpha\beta,\delta\epsilon} = \gamma_{\alpha\beta,\epsilon\delta}, \qquad \text{etc.}$$

We may rewrite the r.h.s. of (5.1.14) in terms of curvature expressions. Namely, if $R^X_{\alpha\beta\gamma\delta}$ and R^Y_{ijkl} are the curvature tensors of X and Y, resp., and if

$$R^X_{\alpha\beta} = \gamma^{\delta\epsilon}R^X_{\alpha\delta\beta\epsilon} = R^X_{\alpha\delta\beta\delta}$$

denotes the Ricci tensor of X, we have in the conventions of J. Jost, Riemannian Geometry amd Geometric Analysis, Springer 1995,

$$R^Y_{jkil} = \frac{1}{2}(g_{jl,ik} + g_{ik,jl} - g_{ij,kl} - g_{kl,ij}) \tag{5.1.15}$$

$$R^X_{\alpha\beta} = \frac{1}{2}(\gamma_{\alpha\delta,\beta\delta} + \gamma_{\beta\delta,\alpha\delta} - \gamma_{\alpha\beta,\delta\delta} - \gamma_{\delta\delta,\alpha\beta}) \tag{5.1.16}$$

(of course, these formulae are only valid in Riemannian normal coordinates.)
From (5.1.14)–(5.1.16), we obtain

$$\Delta\left(\gamma^{\alpha\beta}(x)g_{ij}(f(x))f^i_{x^\alpha}f^j_{x^\beta}\right) = \tag{5.1.17}$$

$$2f^i_{x^\alpha x^\delta}f^i_{x^\alpha x^\delta} + 2R^X_{\alpha\beta}f^i_{x^\alpha}f^i_{x^\beta} - 2R^Y_{jkil}f^j_{x^\alpha}f^k_{x^\delta}f^i_{x^\alpha}f^l_{x^\delta}.$$

This is the desired formula in Riemannian normal coordinates. In order to rewrite it in invariant notation, we observe that in our coordinates $e_\alpha = \frac{\partial}{\partial x^\alpha}, \alpha = 1, \ldots m = \dim X$, constitutes an orthonormal basis of T_xX. We obtain,

$$\frac{1}{2}\Delta\|df\|^2 = \|\nabla df\|^2 + \langle df(Ric^X(e_\alpha)), df(e_\alpha)\rangle_{f^{-1}TY} \tag{5.1.18}$$

$$- \langle R^Y(df(e_\alpha), df(e_\beta))df(e_\beta), df(e_\alpha)\rangle_{f^{-1}TY},$$

where Ric^X denotes the Ricci tensor of X.
In the sequel, of course, either (5.1.17) or (5.1.18) may be employed to reach our subsequent conclusions.

In order to seee how these formulae work, we first consider the case where Y is flat, i.e. $R^Y \equiv 0$, although the consequences in that case will have little to do with nonpositive curvature.

Let thus X be a compact Riemannian manifold. We let

$$p := b_1(X)$$

be the first Betti number of X. Thus, p is the dimension of the space of closed one-forms on X. Integrating a basis of this space over a basis of the space of one-dimensional cycles gives the Albanese map

$$f : X \to T^p$$

to a p-dimensional Euclidean torus. If we use the Hodge theorem and choose our one-forms to be harmonic, then it is not hard to check that f becomes a harmonic map.

Theorem 5.1.1 (Bochner): *Let X be a compact Riemannian manifold of nonnegative Ricci curvature. Then every harmonic one-form is parallel (i.e. $\nabla df \equiv 0$), and*

$$b_1(X) \le m := \dim X.$$

If the Ricci curvature of X is in addition positive at one point at least, then all harmonic one-forms vanish; i.e.

$$b_1(X) = 0.$$

Proof: We integrate (5.1.18) to obtain from the divergence theorem that

$$0 = \frac{1}{2} \int_X \Delta \|df(x)\|^2 \, \mathrm{dvol}(x) \quad = \quad \int_X \|\nabla df(x)\|^2 \, \mathrm{dvol}(x) \qquad (5.1.19)$$

$$+ \int_X \langle df(Ric^X(e_\alpha)), df(e_\alpha) \rangle \, \mathrm{dvol}(x).$$

Since both integrands on the r.h.s. are pointwise nonnegative, they both must be identically zero, i.e.

$$\nabla df(x) \equiv 0 \qquad (5.1.20)$$

$$\langle df(Ric^X(e_\alpha)), df(e_\alpha) \rangle \equiv 0. \qquad (5.1.21)$$

(5.1.20) means that df is parallel. Thus, every harmonic one-form is parallel, and consequently any such harmonic one-form is determined by its value at one given point $x \in X$. In particular, the dimension of the space of harmonic one-forms, i.e. $b_1(X)$, is at most m.

If the Ricci curvature is positive at x, then (5.1.20) implies that

$$df(x) = 0.$$

Since df is parallel, it must therefore vanish identically. Thus, in that case, all harmonic one-forms vanish, and $b_1(X) = 0$.   □

We may perform the same analysis if $f : X \to Y$ is a harmonic map into a Riemannian manifold of nonpositive sectional curvature.

Theorem 5.1.2 (Eells-Sampson): *Let $f : X \to Y$ be harmonic, where X is a compact Riemannian manifold of nonnegative Ricci curvature, Y a Riemannian manifold of nonpositive sectional curvature.*
Then f is totally geodesic, i.e. $\nabla df \equiv 0$, and $\|df\| \equiv const.$
If the Ricci curvature of X is in addition positive at least at one point, then f is constant. If the sectional curvature of Y is negative, then f (is constant or) maps X onto a closed geodesic.

Proof: Again, we integrate (5.1.18) to obtain

$$0 = \frac{1}{2} \int_X \Delta \|df(x)\|^2 \, \mathrm{dvol}(x) \; = \; \int_X \|\nabla df(x)\|^2 \, \mathrm{dvol}(x) \tag{5.1.22}$$

$$+ \int_X \langle df(Ric^X(e_\alpha)), df(e_\alpha) \rangle \, \mathrm{dvol}(x)$$

$$+ \int_X -\langle R^Y(df(e_\alpha)), df(e_\beta))df(e_\beta), df(e_\alpha) \rangle \, \mathrm{dvol}(x).$$

By our assumptions, all three integrands on the r.h.s. of (5.1.22) are pointwise nonnegative, and therefore, they must all vanish identically. From the vanishing of the two first ones, we obtain the analogous conclusion as in the proof of Theorem 5.1.1. If the sectional curvature of Y is negative, then

$$\langle R^Y(df(e_\alpha)), df(e_\beta))df(e_\beta), df(e_\alpha) \rangle \equiv 0$$

implies that $df(e_\alpha)$ and $df(e_\beta)$ are always linearly dependent. Thus, the image of X under f is at most one-dimensional. Since f is totally geodesic, the image must therefore be a closed geodesic (possibly trivial).
Finally, since

$$\Delta \|df(x)\|^2 \geq 0$$

and

$$\int \Delta \|df(x)\|^2 \, \mathrm{dvol}(x) = 0,$$

$\|df(x)\|$ must be constant. $\square$

Remark: The Bochner formula and its consequences remain true for ρ-equivariant harmonic maps $f : M \to N$, where

$$\rho : \Gamma \to I(N)$$

is a homomorphism for a discrete subgroup of $I(M)$ with compact quotient M/Γ, provided of course that the Riemannian manifolds M and N satisfy the necessary curvature conditions.

5.2 A Matsushima type formula for harmonic maps

For the applications of the preceding Bochner type formula, we needed to assume that X has nonnegative Ricci curvature. The idea in the present § will be to refine (5.1.18) so as to balance the Ricci curvature term against the other two terms in such a way that we still get definite signs. This strategy will work in particular when the domain is a locally symmetric space of noncompact type and rank at least 2.

We start with formula (5.1.17). For simplicity of notation, we shall now use an invariant notation on Y, i.e. write

$$\langle f_{x^\alpha}, f_{x^\alpha}\rangle \quad \text{instead of} \quad f^i_{x^\alpha} f^i_{x^\alpha} \quad \text{etc.}$$

We continue to employ Riemannian normal coordinates at the point $x \in X$ under consideration. (5.1.17) thus is

$$\frac{1}{2}\Delta\langle f_{x^\alpha}, f_{x^\alpha}\rangle = \langle f_{x^\alpha x^\delta}, f_{x^\alpha \dot{x}^\delta}\rangle + R^X_{\alpha\beta}\langle f_{x^\alpha}, f_{x^\beta}\rangle - \langle R^Y(f_{x^\alpha}, f_{x^\beta})f_{x^\beta}, f_{x^\alpha}\rangle. \quad (5.2.1)$$

Letting ∇ denote the Levi-Civita connection in $T^*X \otimes f^{-1}TY$ induced by the Riemannian metrics of X and Y, we obtain

$$\left(\nabla_{\frac{\partial}{\partial x^\gamma}}\nabla_{\frac{\partial}{\partial x^\delta}} - \nabla_{\frac{\partial}{\partial x^\delta}}\nabla_{\frac{\partial}{\partial x^\gamma}}\right)(f_{x^\alpha}dx^\alpha) = -R^X_{\alpha\beta\gamma\delta}f_{x^\beta}dx^\alpha + R^Y(f_{x^\gamma}, f_{x^\delta})f_{x^\alpha}dx^\alpha$$
$$(5.2.2)$$

(the negative sign in front of the 1ˢᵗ term on the r.h.s. comes from the fact that the curvature of the cotangent bundle T^*X is the negative of the curvature of the tangent bundle TX, i.e. of X). (5.2.2) yields

$$\langle\left(\nabla_{\frac{\partial}{\partial x^\gamma}}\nabla_{\frac{\partial}{\partial x^\delta}} - \nabla_{\frac{\partial}{\partial x^\delta}}\nabla_{\frac{\partial}{\partial x^\gamma}}\right)f_{x^\alpha}dx^\alpha, \left(\nabla_{\frac{\partial}{\partial x^\gamma}}\nabla_{\frac{\partial}{\partial x^\delta}} - \nabla_{\frac{\partial}{\partial x^\delta}}\nabla_{\frac{\partial}{\partial x^\gamma}}\right)f_{x^\beta}dx^\beta\rangle_{T^*X\otimes f^{-1}TY}$$
$$= R^X_{\beta\alpha\gamma\delta}R^X_{\eta\alpha\gamma\delta}\langle f_{x^\beta}, f_{x^\eta}\rangle_{f^{-1}TY} + \langle R^Y(f_{x^\gamma}, f_{x^\delta})f_{x^\alpha}, R^Y(f_{x^\gamma}, f_{x^\delta})f_{x^\alpha}\rangle_{f^{-1}TY}$$
$$- 2R^X_{\alpha\beta\gamma\delta}\langle R(f_{x^\gamma}, f_{x^\delta})f_{x^\beta}, f_{x^\alpha}\rangle_{f^{-1}TY} \quad (5.2.3)$$

where $\langle\cdot, \cdot\rangle_{T^*X\otimes f^{-1}TY}$ in the first line denotes the induced scalar product in $T^*X\otimes f^{-1}TY$. (5.2.2) also yields

$$\langle\left(\nabla_{\frac{\partial}{\partial x^\gamma}}\nabla_{\frac{\partial}{\partial x^\delta}} - \nabla_{\frac{\partial}{\partial x^\delta}}\nabla_{\frac{\partial}{\partial x^\gamma}}\right)f_{x^\alpha}dx^\alpha, \left(\nabla_{\frac{\partial}{\partial x^\gamma}}\nabla_{\frac{\partial}{\partial x^\delta}} - \nabla_{\frac{\partial}{\partial x^\delta}}\nabla_{\frac{\partial}{\partial x^\gamma}}\right)f_{x^\eta}dx^\eta\rangle_{T^*X\otimes f^{-1}TY}$$
$$= -\langle R^X_{\beta\alpha\gamma\delta}f_{x^\beta}dx^\alpha, \left(\nabla_{\frac{\partial}{\partial x^\gamma}}\nabla_{\frac{\partial}{\partial x^\delta}} - \nabla_{\frac{\partial}{\partial x^\delta}}\nabla_{\frac{\partial}{\partial x^\gamma}}\right)f_{x^\eta}dx^\eta, \rangle_{T^*X\otimes f^{-1}TY}$$
$$+ \langle R^Y(f_{x^\gamma}, f_{x^\delta})f_{x^\alpha}dx^\alpha, \left(\nabla_{\frac{\partial}{\partial x^\gamma}}\nabla_{\frac{\partial}{\partial x^\delta}} - \nabla_{\frac{\partial}{\partial x^\delta}}\nabla_{\frac{\partial}{\partial x^\gamma}}\right)f_{x^\eta}dx^\eta, \rangle_{T^*X\otimes f^{-1}TY}$$
$$= -2\langle R^X_{\beta\alpha\gamma\delta}f_{x^\beta}dx^\alpha, \nabla_{\frac{\partial}{\partial x^\gamma}}\nabla_{\frac{\partial}{\partial x^\delta}}f_{x^\eta}dx^\eta\rangle_{T^*X\otimes f^{-1}TY}$$
$$+ \langle R^Y(f_{x^\gamma}, f_{x^\delta})f_{x^\alpha}, R^Y(f_{x^\gamma}, f_{x^\delta})f_{x^\alpha}\rangle_{f^{-1}TY}$$
$$- R^X_{\alpha\beta\gamma\delta}\langle R^Y(f_{x^\gamma}, f_{x^\delta})f_{x^\beta}, f_{x^\alpha}\rangle_{f^{-1}TY} \quad (5.2.4)$$

using the skew symmetry $R^X_{\beta\alpha\gamma\delta} = -R^X_{\beta\alpha\delta\gamma}$ and once more (5.2.2).

We integrate (5.2.4) over X and integrate the 1st term of the r.h.s. by parts to get

$$\int_X \left\| \left(\nabla_{\frac{\partial}{\partial x^\gamma}} \nabla_{\frac{\partial}{\partial x^\delta}} - \nabla_{\frac{\partial}{\partial x^\delta}} \nabla_{\frac{\partial}{\partial x^\gamma}} \right) f_{x^\alpha} dx^\alpha \right\|^2 = \tag{5.2.5}$$

$$2 \int_X \langle \frac{\partial}{\partial \gamma} \left(R^X_{\beta\alpha\gamma\delta} f_{x^\beta} \right), f_{x^\alpha x^\delta} \rangle + \int_X \langle R^Y (f_{x^\gamma}, f_{x^\delta}) f_{x^\alpha}, R^Y (f_{x^\gamma}, f_{x^\delta}) f_{x^\alpha} \rangle$$

$$- \int_X R^X_{\alpha\beta\gamma\delta} \langle R^Y (f_{x^\gamma}, f_{x^\delta}) f_{x^\beta}, f_{x^\alpha} \rangle.$$

Now

$$\frac{\partial}{\partial \gamma} \left(R^X_{\beta\alpha\gamma\delta} f_{x^\beta} \right) = R^X_{\beta\alpha\gamma\delta} f_{x^\beta x^\gamma} + R^X_{\beta\alpha\gamma\delta,\gamma} f_{x^\beta}. \tag{5.2.6}$$

By the Bianchi identity

$$R^X_{\beta\alpha\gamma\delta,\gamma} = R^X_{\delta\gamma\alpha\gamma,\beta} - R^X_{\delta\gamma\beta\gamma,\alpha} = R^X_{\delta\alpha,\beta} - R^X_{\delta\beta,\alpha}. \tag{5.2.7}$$

If the metric of X is an Einstein metric, then the Ricci tensor is a constant multiple of the metric tensor

$$R_{\alpha\beta} = c\gamma_{\alpha\beta},$$

and consequently, $R_{\alpha\beta,\delta} = 0$ in our Riemannian normal coordinates.

Thus, for an Einstein manifold X,

$$\int_X \left\| \left(\nabla_{\frac{\partial}{\partial x^\gamma}} \nabla_{\frac{\partial}{\partial x^\delta}} - \nabla_{\frac{\partial}{\partial x^\delta}} \nabla_{\frac{\partial}{\partial x^\gamma}} \right) f_{x^\alpha} dx^\alpha \right\|^2 = \tag{5.2.8}$$

$$-2 \int_X R^X_{\alpha\beta\gamma\delta} \langle f_{x^\alpha x^\delta}, f_{x^\beta x^\gamma} \rangle + \int_X \langle R^Y (f_{x^\gamma}, f_{x^\delta}) f_{x^\alpha}, R^Y (f_{x^\gamma}, f_{x^\delta}) f_{x^\alpha} \rangle$$

$$- \int_X R^X_{\alpha\beta\gamma\delta} \langle R^Y (f_{x^\gamma}, f_{x^\delta}) f_{x^\beta}, f_{x^\alpha} \rangle.$$

We now integrate also (5.2.3) and equate the resulting r.h.s. with the r.h.s. of (5.2.8), and we add (5.2.1) multiplied by a real constant λ to get

Lemma 5.2.1 (Jost-Yau): Let $f : X \to Y$ be a harmonic map between Riemannian manifolds, and assume that X is a compact Einstein manifold. Let $\lambda \in \mathbb{R}$. Then

$$\lambda \int_X \langle f_{x^\alpha x^\beta}, f_{x^\alpha x^\beta} \rangle + 2 \int_X R^X_{\alpha\beta\gamma\delta} \langle f_{x^\alpha x^\delta}, f_{x^\beta x^\gamma} \rangle = \tag{5.2.9}$$

$$-\lambda \int_X R^X_{\alpha\beta} \langle f_{x^\alpha}, f_{x^\beta} \rangle - \int_X R^X_{\alpha\beta\gamma\delta} R^X_{\eta\beta\gamma\delta} \langle f_{x^\alpha}, f_{x^\eta} \rangle$$

$$+ \lambda \int_X \langle R^Y (f_{x^\alpha}, f_{x^\beta}) f_{x^\beta}, f_{x^\alpha} \rangle + \int_X R^X_{\alpha\beta\gamma\delta} \langle R^Y (f_{x^\gamma}, f_{x^\delta}) f_{x^\beta}, f_{x^\alpha} \rangle.$$

The first Bochner type identity different from (5.1.18) for harmonic maps was found by Siu [Si1]. Other such formulas are due to Sampson [Sa] and Corlette [Co]. A similar formula as in (5.2.9) was obtained by Mok-Siu-Yeung [MSY], if X is locally symmetric. For the case where $Y = \mathbb{R}$ (or Y is flat), such a formula was already found by Matsushima [Mt]. As in § 5.1, the formula remains true for ρ-equivariant harmonic maps.

As in the previous §, we first apply our formula to the case of a flat image so that the two last terms in the formula disappear. We shall show

Theorem 5.2.1: *Let $X = \Gamma\backslash\overset{G}{/}K$ be a compact, locally irreducible, locally symmetric space of noncompact type, of* rank $(G/K) \geq 2$, *and let T be a flat manifold. Then any harmonic map*

$$f : X \to T$$

is constant.

Corollary 5.2.1 (Matsushima): *Let $X = \Gamma\backslash\overset{G}{/}K$ be a compact, locally irreducible, locally symmetric space of noncompact type, of* rank $(G/K) \geq 2$. *Then the first Betti number of X vanishes,*

$$b_1(X) = 0.$$

Of course, Corollary 5.2.1 is derived from Theorem 5.2.1 in the same manner as Bochner's theorem 5.1.1 was obtained in § 5.1. We therefore turn to the

Proof of Theorem 5.2.1: Since the target is flat, (5.2.9) becomes

$$\lambda \int_X f_{x^\alpha x^\beta} f_{x^\alpha x^\beta} + 2 \int_X R^X_{\alpha\beta\gamma\delta} f_{x^\alpha x^\delta} f_{x^\beta x^\gamma} = \tag{5.2.10}$$

$$-\lambda \int_X R^X_{\alpha\beta} f_{x^\alpha} f_{x^\beta} - \int_X R^X_{\alpha\beta\gamma\delta} R^X_{\eta\beta\gamma\delta} f_{x^\alpha} f_{x^\eta}$$

(here $f_{x^\alpha} f_{x^\eta}$ etc. just stands for the scalar product of Euclidean vectors; in fact, it suffices to assume that f is scalar valued.)

The strategy now is to find a suitable value for λ so that the l.h.s. of (5.2.10) is positive unless all $f_{x^\alpha x^\beta} \equiv 0$ while the r.h.s. is nonpositive, unless all $f_{x^\alpha} \equiv 0$. Consequently we need to find a value λ for which

$$2R^X_{\alpha\beta\gamma\delta} f_{x^\alpha x^\delta} f_{x^\beta x^\gamma} \geq -\lambda f_{x^\alpha x^\beta} f_{x^\alpha x^\beta} \tag{5.2.11}$$

and $\qquad R^X_{\alpha\beta\gamma\delta} R^X_{\eta\beta\gamma\delta} f_{x^\alpha} f_{x^\eta} \geq -\lambda R^X_{\alpha\eta} f_{x^\alpha} f_{x^\eta}. \tag{5.2.12}$

Concerning (5.2.11), Calabi-Vesentini [CV] and Borel [Bo] in the Hermitian case (see [Mt], § 9 for a table) and Kaneyuki-Nagano [KN] for the remaining cases computed the optimal λ_1 satisfying

$$R^X_{\alpha\beta\gamma\delta} A_{\alpha\delta} A_{\beta\gamma} \geq -\lambda_1 A_{\alpha\beta} A_{\alpha\beta} \tag{5.2.13}$$

for symmetric, traceless matrices $(A_{\alpha\beta})_{\alpha,\beta=1,\dots,m}$.

We shall therefore need to relate that constant λ_1 to the optimal parameter value in (5.2.12). As in § 1.1, we let $\mathfrak{g}$ be the Lie algebra of G, $\mathfrak{k}$ the one of K, and we use the decomposition

$$\mathfrak{g} = \mathfrak{k} \oplus \mathfrak{p}$$

with associated bases $X_1, \ldots, X_m$ of $\mathfrak{p}$, $X_{m+1}, \ldots, X_{m+k}$ of $\mathfrak{k}$, satisfying

$$B(X_\alpha, X_\beta) = \delta_{\alpha\beta} \quad \text{for } \alpha, \beta = 1, \ldots, m \tag{5.2.14}$$
$$B(X_\lambda, X_\mu) = -\delta_{\lambda\mu} \quad \text{for } \lambda, \mu = m+1, \ldots, m+k \tag{5.2.15}$$

for the Killing form B of $\mathfrak{g}$. Such a basis of $\mathfrak{g}$ will be called orthonormal. We also let $c_{\alpha\beta}^\lambda$ be the structure constants of the symmetric space G/K, i.e.

$$[X_\alpha, X_\beta] = c_{\alpha\beta}^\lambda X_\lambda \quad \text{for } \alpha, \beta = 1, \ldots, m \tag{5.2.16}$$

(note that $[\mathfrak{p}, \mathfrak{p}] \subset \mathfrak{k}$, and so λ here only ranges from $m+1$ to $m+k$).

One computes

$$c_{\alpha\beta}^\lambda c_{\gamma\beta}^\lambda = \frac{1}{2}\delta_{\alpha\gamma} \quad \text{for } \alpha, \gamma = 1, \ldots, m. \tag{5.2.17}$$

We recall the formula (1.1.11) for the curvature tensor of G/K and X (omitting the superscript X from now on):

$$R(X,Y)Z = -[[X,Y],Z]. \tag{5.2.18}$$

Consequently

$$
\begin{aligned}
R_{\alpha\beta\gamma\delta} &= B(R(X_\alpha, X_\beta)X_\delta, X_\gamma) \\
&= -B([[X_\alpha, X_\beta], X_\delta], X_\gamma) \\
&= B([X_\alpha, X_\beta], [X_\gamma, X_\delta]) \quad \text{as in (1.1.12)} \\
&= -c_{\alpha\beta}^\lambda c_{\gamma\delta}^\lambda \quad \text{from (5.2.16), (5.2.15).}
\end{aligned}
\tag{5.2.19}
$$

Recalling (5.2.17), we get for the Ricci tensor

$$R_{\alpha\beta} = -\frac{1}{2}\delta_{\alpha\beta}. \tag{5.2.20}$$

We now need the decomposition

$$\mathfrak{k} = \mathfrak{z} \oplus \mathfrak{k}_1 \oplus \ldots \oplus \mathfrak{k}_l \tag{5.2.21}$$

where $\mathfrak{z}$ is the (possibly trivial) center of $\mathfrak{k}$, and $\mathfrak{k}_1, \ldots, \mathfrak{k}_l$ are the simple ideals (in fact, $l = 1$ or 2). We choose our basis vectors X_λ of $\mathfrak{k}$ so that each X_λ is contained in one of the summands in (5.2.4). One then has

$$c_{\alpha\beta}^\lambda c_{\alpha\beta}^\mu = a_i \delta_{\lambda\mu} \quad \text{for } X_\lambda \in \mathfrak{k}_i, \ i = 1, \ldots, l, \tag{5.2.22}$$

with
$$0 < a_i < 1,$$
and
$$c_{\alpha\beta}^{\lambda} c_{\alpha\beta}^{\mu} = \delta_{\lambda\mu} \quad \text{for } X_\lambda \in \mathfrak{z}. \tag{5.2.23}$$

In order to have a uniform notation, we put $\mathfrak{k}_0 := \mathfrak{z}$, and $a_0 = 1$ for $\mathfrak{z} \neq \{0\}$, $a_0 = 0$ for $\mathfrak{z} = \{0\}$.

Finally, for $i = 0, \ldots, l$
$$\sum_{X_\lambda \in \mathfrak{k}_i} c_{\alpha\beta}^{\lambda} c_{\eta\beta}^{\lambda} = b_i \delta_{\alpha\eta}, \tag{5.2.24}$$

with
$$\sum_{i=0}^{l} b_i = \frac{1}{2} \quad \text{by (5.2.17).} \tag{5.2.25}$$

We then have
$$\begin{aligned}
R_{\alpha\beta\gamma\delta} R_{\eta\beta\gamma\delta} f_{x^\alpha} f_{x^\eta} &= c_{\alpha\beta}^{\lambda} c_{\gamma\delta}^{\lambda} c_{\eta\beta}^{\mu} c_{\gamma\delta}^{\mu} f_{x^\alpha} f_{x^\eta} \quad \text{by (5.2.19)} \\
&= a_i b_i f_{x^\alpha} f_{x^\alpha} \quad \text{by (5.2.22), (5.2.24)} \\
&= -2 a_i b_i R_{\alpha\eta} f_{x^\alpha} f_{x^\eta}. \tag{5.2.26}
\end{aligned}$$

With $\mu := 2 a_i b_i$, we then get equality in (5.2.12) for the value $\lambda = \mu$. One then verifies by computations of Matsushima [Mt] and Kaneyuki-Nagano [KN]
$$\mu > -2\lambda_1 \quad \text{in case } \operatorname{rank}(G/K) \geq 2. \tag{5.2.27}$$

Therefore, we may choose any λ with
$$-2\lambda_1 < \lambda < \mu$$

and get
$$2 R_{\alpha\beta\gamma\delta} f_{x^\alpha x^\delta} f_{x^\beta x^\gamma} > -\lambda f_{x^\alpha x^\beta} f_{x^\alpha x^\beta} \quad \text{unless all } f_{x^\alpha x^\beta} \equiv 0$$

and
$$R_{\alpha\beta\gamma\delta} R_{\eta\beta\gamma\delta} f_{x^\alpha} f_{x^\eta} > -\lambda R_{\alpha\eta} f_{x^\alpha} f_{x^\eta} \quad \text{unless all } f_{x^\alpha} \equiv 0.$$

As explained above, (5.2.10) then allows us to conclude that all $f_{x^\alpha} \equiv 0$, i.e. f is constant. □

Remark: It follows from (5.1.16) that if X is a compact manifold with negative or positive definite Ricci tensor, and if $f : X \to T$ is a totally geodesic map into a flat space, then f is constant. Namely, integrating (5.1.17) for a totally geodesic f into a flat target, one gets
$$\int R_{\alpha\beta}^X f_{x^\alpha} f_{x^\beta} = 0,$$

hence
$$f_{x^\alpha} \equiv 0 \quad \text{for all } \alpha \text{ by definiteness of } R_{\alpha\beta}^X.$$

Therefore, in the preceding proof, it would have been sufficient to conclude that f is totally geodesic, i.e. $f_{x^\alpha x^\beta} \equiv 0$ for all α, β.

5.3   Geometric superrigidity

We may now realize the goal stated in § 1.3, namely to use harmonic maps to prove some of the superrigidity theorems of Margulis. We start with the archimedean case:

Theroem 5.3.1 (Margulis): *Let $G/_K$ be an irreducible symmetric space of noncompact type and* $\operatorname{rank}(G/_K) \geq 2$, *$\Gamma$ a cocompact lattice in G. Let H be a semisimple noncompact Lie group (without compact factors) with trivial center, $\rho : \Gamma \to H$ a homomorphism with Zariski dense image. Then ρ extends to a homomorphism from G to H.*

Proof: We identify H as the isometry group $I(N)$ of a symmetric space N of noncompact type, and from our general existence theorem, we obtain a ρ-equivariant harmonic map

$$\tilde{f} : G/_K \to N;$$

on the level of quotients, we have a harmonic map

$$f : X \left(= \Gamma \backslash G/_K \right) \to Y \left(= N/_{\rho(\Gamma)} \right),$$

and, as before, we assume for simplicity that X and Y are compact smooth manifolds, in order not to have to work on the level of universal coverings.

We wish to use Lemma 5.2.1 to show that f is totally geodesic. As explained in § 1.3, one may then easily obtain the homomorphism from G to H. Again, the game is to find a suitable value for λ, so that

$$2R^X_{\alpha\beta\gamma\delta}\langle f_{x^\alpha x^\delta}, f_{x^\beta x^\gamma}\rangle \geq -\lambda\langle f_{x^\alpha x^\beta}, f_{x^\alpha x^\beta}\rangle \tag{5.3.1}$$

(wih strict inequality unless all $f_{x^\alpha x^\beta} \equiv 0$)

$$R^X_{\alpha\beta\gamma\delta}R^X_{\eta\beta\gamma\delta}\langle f_{x^\alpha}, f_{x^\eta}\rangle \geq -\lambda R_{\alpha\beta}\langle f_{x^\alpha}, f_{x^\eta}\rangle \tag{5.3.2}$$

$$R^X_{\alpha\beta\gamma\delta}\langle R^Y(f_{x^\gamma}, f_{x^\delta})f_{x^\beta}, f_{x^\alpha}\rangle \leq -\lambda\langle R^Y(f_{x^\alpha}, f_{x^\beta})f_{x^\beta}, f_{x^\alpha}\rangle. \tag{5.3.3}$$

In the sequel, we shall omit the superscripts X and Y for the curvature tensors. As the inequalities in (5.3.2) and (5.3.3) go in opposite directions, the estimates now need to be tighter than the ones in § 5.2. In fact, one has to show that (5.3.3) holds with $\lambda = \mu = 2a_ib_i$ (cf. (5.2.26)). On the other hand, we only need the following nonpositivity of the curvature tensor of Y:
Put

$$P_{\alpha\beta\gamma\delta} := -\langle R(f_{x^\alpha}, f_{x^\beta})f_{x^\delta}, f_{x^\gamma}\rangle.$$

Then P is positive semidefinite in the sense that

$$2P_{\alpha\beta\gamma\delta} \leq P_{\alpha\beta\alpha\beta} + P_{\gamma\delta\gamma\delta} \quad \text{for all } \alpha, \beta, \gamma, \delta. \tag{5.3.4}$$

Since coming from a curvature tensor, P also satisfies the identities

$$P_{\alpha\beta\gamma\delta} = -P_{\beta\alpha\gamma\delta} \tag{5.3.5}$$

$$P_{\alpha\beta\gamma\delta} = P_{\gamma\delta\alpha\beta} \tag{5.3.6}$$

$$P_{\alpha\beta\gamma\delta} + P_{\alpha\gamma\delta\beta} + P_{\alpha\delta\beta\gamma} = 0 \quad \text{(Bianchi identity)} \tag{5.3.7}$$

for all α, β, γ, δ.

The verification of (5.3.3) with $\lambda = \mu$ was obtained in [JY2] by a case-by-case study, essentially according to the number of summands in (5.2.21) (there are at most three such summands), and by a somewhat different case-by-case analysis in [MSY]. Here, we only treat the simplest case, namely where $\mathfrak{k}$ is simple. Then

$$c_{\alpha\beta}^{\lambda} c_{\alpha\beta}^{\mu} = a\delta_{\lambda\mu} \quad \text{for all } \lambda, \mu \tag{5.3.8}$$

with $0 < a < 1$. (5.2.26) then becomes, recalling (5.2.25),

$$R_{\alpha\beta\gamma\delta} R_{\eta\beta\gamma\delta} \langle f_{x^\alpha}, f_{x^\beta} \rangle = c_{\alpha\beta}^{\lambda} c_{\gamma\delta}^{\lambda} c_{\eta\beta}^{\mu} c_{\gamma\delta}^{\mu} \langle f_{x^\alpha}, f_{x^\eta} \rangle \tag{5.3.9}$$

$$= -a R_{\alpha\eta} \langle f_{x^\alpha}, f_{x^\eta} \rangle.$$

Also,

$$-R_{\alpha\beta\gamma\delta} P_{\alpha\beta\gamma\delta} = c_{\alpha\beta}^{\lambda} c_{\gamma\delta}^{\lambda} P_{\alpha\beta\gamma\delta}$$

$$\leq a P_{\alpha\beta\alpha\beta} \; ; \tag{5.3.10}$$

the last inequality follows from the Schwarz inequality, since the $\left(c_{\alpha\beta}^{\lambda}\right)_{\alpha,\beta=1,\ldots,m}$ for $\lambda = m+1, \ldots, m+k$, are pairwise orthogonal and of the same squared norm a by (5.2.8).

This is (5.3.3) with $\lambda = \mu = a$.

By (5.2.27), if we insert $\lambda = \mu$ in (5.2.9), we obtain that f is totally geodesic, i.e. $f_{x^\alpha x^\beta} \equiv 0$ for all α, β (in our normal coordinates).

This finishes the sketch of the proof.   □

We now turn to the nonarchimedean case:

Theorem 5.3.2 (Margulis): *Let G/K and Γ be as in the preceding theorem. Let $\rho : \Gamma \to Sl\,(n, \mathbb{Q}_p)$ be a (reductive) homomorphism, for some $n \in \mathbb{N}$ and some prime p. Then $\rho(\Gamma)$ has compact closure in $Sl\,(n, \mathbb{Q}_p)$.*

Again, we shall only provide a sketch of the proof:

As explained in § 1.3, $Sl\,(n, \mathbb{Q}_p)$ operates on some Euclidean Bruhat-Tits building N, which is a special case of a global Alexandrov NPC space. According to the general Theorem 4.2.1, there exists a ρ-equivariant harmonic map

$$f : G/K \to N.$$

N is not a smooth manifold anymore, but by the analysis of Gromov-Schoen [GS], the singularities of f can be sufficiently well controlled so that Lemma 5.2.1 remains valid (unfortunately, we have to omit all details of this point). For this reason, the analysis of the preceding proof can be carried over to conclude that f is totally geodesic. Since N is piecewise flat, one may then use the remark after the proof of Theorem 5.2.1 to show that f is constant. Thus, $f \equiv q$ for some $q \in N$, and q therefore is a fixed point of $\rho(\Gamma)$ by ρ-equivariance of f. This, however, means that $\rho(\Gamma)$ is contained in the isotropy group of q, a compact subgroup of $Sl\,(n, \mathbb{Q}_p)$. $\square$

Remarks:

1) As mentioned, the proof of Theorem 5.3.2 needs regularity results for generalized harmonic maps which are not given here. Gromov-Schoen [GS] obtained precise estimates if the domain is a Riemannian manifold and the image is a Euclidean Bruhat-Tits building. Korevaar-Schoen [KS] showed Lipschitz continuity if the domain is a Riemannian manifold and the image is an Alexandrov NPC space. Jost [Jo7] showed Hölder continuity if the domain satisfies a Poincaré type inequality and a ball doubling property (the measure of a ball of radius $2r$ has to be controlled by a constant times the measure of the ball of radius r, for $0 < r \leq R_0$ for some fixed $R_0 > 0$; in the Riemannian case, such a ball doubling property follows from a lower bound for the Ricci curvature), and the image again is an Alexandrov NPC space. The proof extends to the case of a uniform Busemann NPC space.

2) The preceding results extend to the cases $G/K = Sp(p,1)/Sp(p) \times Sp(1)$ and the hyperbolic Cayley plane, and one may actually use the same kind of analysis. The analogue of Theorem 5.3.1 in this situation is due to Corlette [Co], the analogue of Theorem 5.3.2 to Gromov-Schoen [GS].

We return to the archimedean case and observe the following

Corollary 5.3.1: *Let* $X = \Gamma \backslash G/K$ *be a compact, locally irreducible, locally symmetric space of noncompact type and* rank $(G/K) \geq 2$. *Let* $g : X \to X$ *be continuous. Then*

$$|\deg g| \leq 1.$$

If $\deg g = \pm 1$, *then* g *is homotopic to an isometry, if* $\deg g = 0$, *to a constant map.*

Proof: g is homotopic to a harmonic map f. By the proof of Theorem 5.3.1, f is totally geodesic.

Either $\|df(e)\|$, e a unit tangent vector, then is independent of e, or we may obtain an orthogonal splitting by choosing a minimal (or maximal) direction e and its orthogonal complement (since f is totally geodesic, this does not depend on $x \in X$). Such a nontrivial splitting does not exist for an irreducible X, and thus

$$\|df(e)\| \equiv c_0. \tag{5.3.11}$$

We claim that $c_0 \leq 1$. Since f is totally geodesic, we obtain from (5.2.20) and integration of (5.1.17)

$$-\frac{1}{2}\langle f_{x^\alpha}, f_{x^\alpha}\rangle = R_{\alpha\beta}\langle f_{x^\alpha}, f_{x^\beta}\rangle = \langle R(f_{x^\alpha}, f_{x^\beta})f_{x^\beta}, f_{x^\alpha}\rangle. \tag{5.3.12}$$

We consider the iterates $f^n = f \circ \ldots \circ f$. Since the l.h.s. of (5.3.12) grows quadratically in n, while the r.h.s. grows at the fourth power, we must have $c_0 \leq 1$ indeed. Therefore, also

$$\|d(f^n)(e)\| \leq 1 \quad \text{for all } n \in \mathbb{N} \text{ and all unit tangent vectors } e. \tag{5.3.13}$$

This means that the iterates f^n are equicontinuous, they then have to converge to some limit map f^0, after selection of a subsequence at least. Therefore, f^n must be homotopic to f^0 for large n, and since all f^n are totally geodesic, hence harmonic, uniqueness (Corollary 4.2.1) implies that $f^n = f^0$ for large n. Therefore,

$$\|d(f^n)(e)\| \equiv 0 \text{ or } 1$$

for large, hence for all n. If $\|d(f^n)(e)\| \equiv 0$ for all e, then f^n is constant, while if $\|d(f^n)(e)\| \equiv 1$ for all e with $\|e\| = 1$, then f^n is an isometry.

If f is an isometry, we must have $\deg f = \pm 1$, because if $|\deg f| \geq 2$, again consideration of the iterates f^n (which then would have arbitrarily large degree) would not be compatible with (5.3.12). $\qquad\square$

Bibliography

This bibliography contains only those references that are quoted in the text. A more extensive bibliography can be found in [Ba].

[AS] U. Abresch, V. Schroeder, *Analytic manifolds of nonpositive curvature*, Preprint

[AB] S.B. Alexander, R.L. Bishop, *The Hadamard-Cartan theorem in locally convex metric spaces*, L'Enseign. Math. 36, 1990, p. 309–320

[Ab] W. Abikoff, *The real analytic theory of Teichmüller space*, Springer, Lecture Notes Mathematics 820, Berlin Heidelberg New York, 1980

[A1] S.I. Al'ber, *On n-dimensional problems in the calculus of variations in the large*, Sov. Math. Dokl. 5, 1964, p. 700–704

[A2] S.I. Al'ber, *Spaces of mappings into a manifold with negative curvature*, Sov. Math. Dokl. 9, 1967, p. 6–9

[Ar] S.J. Arakelov, *Families of algebraic curves with fixed degeneracies*, Math. USSR Izv. 5, 1971, p. 1277–1302

[BGS] W. Ballmann, M. Gromov, V. Schroeder, *Manifolds of nonpositive curvature*, Birkhäuser, Progress in Mathematics, Basel Boston Berlin, 1985

[Ba] W. Ballmann, *Lectures on spaces of nonpositive curvature*, Birkhäuser, Basel Boston Berlin, 1995

[BM] M. Burger, S. Mozes, *Finitely presented simple groups and products of trees*, Preprint

[BN] V.N. Berestovskij, I.G. Nikolaev, *Multidimensional generalized Riemannian spaces*, Geometry IV (ed. Y.G. Reshetnyak), Encyclopaedia of Math. Sciences, Vol. 70, Springer, Berlin Heidelberg New York, 1993

[Bo] A. Borel, *On the curvature tensor of the Hermitian symmetric manifolds*, Ann. Math. 71, 1960, p. 508–521

[Br] K.S. Brown, *Buildings*, Springer, Berlin Heidelberg New York, 1989

[Bu] H. Busemann, *The geometry of geodesics*, Academic Press, New York, 1955

[CV] E. Calabi, E. Vesentini, *On compact, locally symmetric Kähler manifolds*, Ann. Math. 71, 1960, p. 472–507

[Co] K. Corlette, *Archimedian superrigidity and hyperbolic geometry*, Ann. Math. 135, 1990, p. 165–182

[DSW] X.Z. Dai, Z.M. Shen, G.F. Wei, *Negative Ricci curvature and isometry group*, Duke Math. Jour. 76, 1994, p. 59–73

[DaM] G. Dal Maso, *An introduction to Γ-convergence*, Birkhäuser, Basel Boston Berlin, 1993

[DM] P. Deligne, D. Mumford, *The irreducibility of the space of curves of given genus*, Publ. Math. IHES 36, 1969, p. 75–110

[DO] K. Diederich, T. Ohsawa, *Harmonic mappings and disc bundles over compact Kähler manifolds*, Publ. Res. Inst. Math. Sci. 21, 1985, p. 819–833

[Do] S.K. Donaldson, *Twisted harmonic maps and the self-duality equations*, Proc. London Math. Soc. 55, 1987, p. 127–131

[Eb] P. Eberlein, *Structure of manifolds of nonpositive curvature*, Springer, Lecture Notes Mathematics 1156, Berlin Heidelberg New York, 1984, p. 86–153

[EL] J. Eells, L. Lemaire, *Selected topics in harmonic maps*, CBMS, Amer. Math. Soc. 50, 1983

[ES] J. Eells, J. Sampson, *Harmonic mappings of Riemannian manifolds*, Amer. Jour. Math. 85, 1964, p. 109–160

[Fa] G. Faltings, *Arakelov's theorem for Abelian varieties*, Inv. math. 73, 1973, p. 337–347

[FJ1] F.T. Farrell, L.E. Jones, *A topological analogue of Mostow's rigidity theorem*, Jour. Amer. Math. Soc. 2, 1989, p. 257–370

[FJ2] F.T. Farrell, L.E. Jones, *Negatively curved manifolds with exotic smooth structures*, Jour. Amer. Math. Soc. 4, 1989, p. 899–908

[FJ3] F.T. Farrell, L.E. Jones, *Rigidity and other topological aspects of compact nonpositively curved manifolds*, Bull. of the AMS 22, 1990, p. 59–64

[FJ4] F.T. Farrell, L.E. Jones, *Nonuniform hyperbolic lattices and exotic smooth structures*, Jour. Diff. Geom. 38, 1993, p. 235–261

[Fr] Frankel, S., *Locally symmetric and rigid factors for complex manifolds via harmonic maps*, Ann. Math. 141, 1995, p. 285–300

[Gr] H. Grauert, *Mordells Vermutung über rationale Punkte auf algebraischen Kurven und Funktionenkörper*, Publ. Math. IHES 25, 1965 p. 363–381

[GR] H. Grauert, H. Reckziegel, *Hermitesche Metriken und normale Familien holomorpher Abbildungen*, Math. Zeit. 89, 1965, p. 108–125

[G1] M. Gromov, *Manifolds of negative curvature*, Jour. Diff. Geom. 13, 1978, p. 223–230

[G2] M. Gromov, *Hyperbolic manifolds, groups and actions*, Riemann surfaces and related topics (I. Kra and B. Maskit, eds.), Ann. Math. Studies 97, Princeton University Press, 1981, p. 183–213

[G3] M. Gromov, *Hyperbolic groups*, Essays in group theory, (S.M. Gersten, ed.), Math. Sci. Res. Inst. Publ. 8, 1987, p. 75–263

[GP] M. Gromov, I. Piatetski-Shapiro, *Non-arithmetic groups in Lobachevsky spaces*, Publ. Math. IHES 66, 1988, p. 93–103

[GS] M. Gromov, R. Schoen, *Harmonic maps into singular spaces and p-adic superrigidity for lattices in groups of rank one*, Publ. Math. IHES 76, 1992, p. 165–246

[GT] M. Gromov, W. Thurston, *Pinching constants for hyperbolic manifolds*, Inv. math. 89, 1987, p. 1–12

[Ha] P. Hartman, *On homotopic harmonic maps*, Can. Jour. Math. 19, 1967, p. 673–687

[He] S. Helgason, *Differential geometry, Lie groups and symmetric spaces*, Academic Press, New York, 1978

[Hn] E. Heintze, *Mannigfaltigkeiten negativer Krümmung*, Habilitationsschrift, Universität Bonn, 1976

[HS] C. Hummel, V. Schroeder, *Cusp closing in rank one symmetric spaces*, Inv. math., 1996, p. 283–307

[Hu] B. Hunt, *A bound on the Euler number for certain Calabi-Yau manifolds*, Jour. reine angewandte Math. 411, 1990, p. 137–170

[IH] H.-C. Im Hof, *Über die Isometriegruppe bei kompakten Mannigfaltigkeiten negativer Krümmung*, Comm. Math. Helv. 48, 1973, p. 14–30

[Jo1] J. Jost, *Existence proofs for harmonic mappings with the help of a maximum principle*, Math. Zeit. 184, 1983, p. 489–496

[Jo2] J. Jost, *Harmonic maps and curvature computations in Teichmüller theory*, Ann. Acad. Sci. Fen., Series A. I. Math. 16, 1991, p. 13–46

[Jo3] J. Jost, *Two dimensional geometric variational problems* , Pure and Applied Mathematics Series, Wiley-Interscience Publ., Chichester, 1991

[Jo4] J. Jost, *Equilibrum maps between metric spaces*, Calc. Var. 2, 1994, p. 173–204

[Jo5] J. Jost, *Convex functionals and generalized harmonic maps into spaces of nonpositive curvature*, Comment. Math. Helv. 70, 1995, p. 659–673

[Jo6] J. Jost, *Riemannian geometry and geometric analysis*, Springer, Berlin Heidelberg New York, 1995

[Jo7] J. Jost, *Generalized Dirichlet forms and harmonic maps*, Calc. Var., 5, 1997, p. 1–19

[Jo8] J. Jost, *Generalized harmonic maps between metric spaces*, in: Geometric Analysis and the Calculus of Variations for Stefan Hildebrandt (J. Jost, ed.), Intern. Press, Boston, 1996, p. 143–174

[JP]   J. Jost, X.-W. Peng, *Group actions, gauge transformations,and the calculus of variations*, Math. Ann. 293, 1992, p. 595–621

[JY1] J. Jost, S.T. Yau, *Harmonic maps and group representations*, Differential geometry, H.B. Lawson and K. Tenenblat, eds., Longman, p. 241–259

[JY2] J. Jost, S.T. Yau, *Harmonic maps and superrigidity*, Differential geometry: partial differential equations on manifolds, Proc. Symp. Pure Math. 54, part I, 1993, p. 245–280

[JY3] J. Jost, S.T. Yau, *Harmonic mappings and algebraic varieties over function fields*, Amer. Jour. of Math. 115, 1993, p. 1197–1227

[KS]   N. Korevaar, R. Schoen, *Sobolev spaces and harmonic maps for metric space targets*, Comm. Anal. Geom. 1, 1993, p. 561–569

[La]   F. Labourie, *Existence d'applications harmoniques tordues à valeurs dans les variétés à courbure négative*, Proc. Amer. Math. Soc. 111, 1991, p. 877–882

[Lo]   J. Lohkamp, *Metrics of negative Ricci curvature*, Ann. Math. 140, 1994, p. 655–683

[Lj]   E. Looijenga, , *A Torelli thorem for Kähler-Einstein K3 surfaces*, Lect. Notes Math. 894, p. 107–112, 1982

[Mk]   V.S. Makarov, *On a certain class of discrete Lobachevsky space groups with infinite fundamental domain of finite measure*, Soviet Math. Dokl. 7, 1966, p. 328–331

[Ma] J. Manin, *Rational points of algebraic curves over function fields*, Izv. Akad. Nauk. SSSR, Ser. Mat. 27, 1963, p. 1395–1440

[Mg1] G.A. Margulis, *Discrete groups of motion of manifolds of nonpositive curvature* Amer. Math. Soc. Translations 190, 1977, p. 33–45

[Mg2] G.A. Margulis, *Discrete subgroups of semisimple Lie groups*, Springer, Ergebnisse der Mathematik und ihrer Grenzgebiete 3/17, Berlin Heidelberg New York, 1991

[Mg3] G.A. Margulis, *Superrigidity for commensurability subgroups and generalized harmonic maps*, Preprint

[Mt] Y. Matsushima, *On the first Betti number of compact quotient spaces of higher dimensional symmetric spaces*, Ann. Math. 75, 1962, p. 312–330

[Mi] J. Milnor, *A note on curvature and fundamental group*, Jour. Diff. Geom., Vol. 2, 1968, p. 1–7

[MSY] N. Mok, Y.T. Siu, S.K. Yeung, *Geometric superrigidity*, Inv. math. 113, 1993, p. 57–84

[Mo] G.D. Mostow, *Strong rigidity of locally symmetric spaces*, Ann. Math. Studies 78, Princeton University Press, 1973

[MS] G.D. Mostow, Y.S. Siu, *A complex Kähler surface of negative curvature not covered by the ball*, Ann. Math. 112, 1980, p. 321–360

[No] J. Noguchi, *Moduli spaces of holomorphic mappings into hyperbolically imbedded complex spaces and locally symmetric spaces*, Inv. math. 93, 1988, p. 15–34

[Ol] A. Yu. Olshanski, *The SQ-universality of hyperbolic groups*, Mat. Sb. (Russ. Akad. Nauk) 186, 1995, p. 1199–1212

[Pa] A.N. Parshin, *Algebraic curves over function fields, I*, Math. USSR Izv. 2, 1968, p. 1145–1170

[Pr] G. Prasad, *Strong rigidity of Q-rank 1 lattices*, Inv. math. 21, 1973, p. 255–286

[Pm] A. Preissmann, *Quelques propriétés des espaces de Riemann*, Comment. Math. Helv. 15, 1942–43, p. 175–216

[Re] Y.G. Reshetnyak, *Inextensible mappings in a space of curvature no greater than K*, Sib. Math. Jour. 9, 1968, p. 683–689

[Ro] H. Royden, *The Ahlfors-Schwarz lemma in several complex variables*, Comment. Math. Helv. 55, 1980, p. 574–558

[Sa]   J. Sampson, *Applications of harmonic maps to Kähler geometry*, Contemp. Math 49, p. 125–134

[Se]   J.-P. Serre, *Trees*, Springer, Berlin Heidelberg New York, 1980

[Si1]   Y.T. Siu, *The complex-analicity of harmonic maps and the strong rigidity of compact Kähler manifolds*, Ann. Math. 112, 1980, p. 73–111

[Si2]   Y.T. Siu, *A simple proof of the surjectivity of the period map of K3 surfaces*, manuscr. math. 35, 1981, p. 225–255

[Si3]   Y.T. Siu, *Curvature of the Weil-Petersson metric in the moduli space of compact Kähler-Einstein manifolds of negative first Chern class* in K. Diederich (ed.): Aspects of Mathematics, vol. 9, p. 261–298, Vieweg

[Sl]   Z. Sela, *The isomorphism problem for hyperbolic groups I*, Ann. Math. 141, 1995, p. 217–283

[Sv]   A.S. Švarc, *A volume invariant of coverings*, Dokl. Akad. Nauk. SSSR 105, 1955, p. 32–40

[To]   A. Todorov, *Applications of the Kähler-Einstein Calabi-Yau metric to moduli of K3 surfaces*, Inv. math. 61, 1980, p. 251–265

[Tr]   A.J. Tromba, *On a natural connection on the space of almost complex structures and the curvature of Teichmüller space with respect to its Weil-Petersson metric*, Manusc. Math. 56, 1986, p. 475–497

[Vi]   E.B. Vinberg, *Hyperbolic reflection groups*, Usp. Math. Nauk. 40, 1985, p. 29–66

[Wo]   S. Wolpert, *Chern forms and the Riemann tensor for the moduli space of curves*, Inv. math. 85, 1986, p. 119–145

[Ya]   S.T. Yau, *On the Ricci curvature of a compact Kähler manifold and the complex Monge-Ampère equation, I*, Comm. Pure Appl. Math. 31, 1978, p. 339–413

[Zh1]   F. Zheng, *Examples of nonpositively curved Kähler manifolds*, Comm. Anal. Geom., to appear

[Zh2]   F. Zheng, *Hirzebruch-Kato surfaces, Deligne-Mostow's construction, and new examples of negatively curved compact Kähler surfaces*, Preprint, 1996

[Zi]   R. Zimmer, *Ergodic theory and semi-simple groups*, Birkhäuser, Monographs in Mathematics, Basel Boston Berlin, 1984

Index

Abelian subalgebra, 16
Abelian subgroup, 4
Abelian subspace, 16
Abelian variety, 21
Abresch, 7
Adams, 76
Ahlfors-Yau-Royden-Schwarz
  lemma, 2
Albanese map, 88
Al'ber, 81, 83
Alexander, 44, 45, 51–53
Alexandrov, 41, 43
Alexandrov NPC inequality, 54
Alexandrov NPC space, 54, 56–58,
  67, 74, 81, 96
algebraic group, 27
algebraic surface, 19
angle, 55
Arakelov, 19
archimedean, 95, 97
arithmetic, 29
arithmetic lattice, 29
arithmeticity, 27, 29
automorphism group, 5

Ballmann, 18
Berestovskij, 55
Bianchi identity, 15, 91, 96
Bishop, 44, 45, 51–53
Bochner, 88, 92
Bochner formula, 89
Borel, 18, 29, 92
Borel density theorem, 28
Brown, 31
Burger, 10

Busemann, 41, 43, 45
Busemann NPC inequality,
  44, 47, 48
Busemann NPC space, 44, 45, 49–
  54, 58, 61, 62, 65–67, 74, 76,
  78, 81–83, 97

Calabi, 92
canonical homotopy, 4
capsule, 50
Cartan, 66
Cauchy polar decomposition
  theorem, 13
center, 93
center of gravity, 65
center of mass, 65, 68, 72
closed ball, 43
closed geodesic, 4, 61
cocompact, 18
cocompact lattice, 95
commensurability subgroup, 81
commensurable, 29, 83
complex hyperbolic space, 29
complexified sectional curvature, 3
condition (C), 62
connectivity assumption, 79
convex, 36, 38, 40, 44, 46, 61, 75
convex hull, 67, 68
convexity, 62, 83
Corlette, 29, 81, 92

Dal Maso, 75
De Giorgi, 74
Diederich, 81
discrete, 18

distance nonincreasing, 3
Donaldson, 81

Eberlein, 18
Eells, 81, 85, 89
Einstein manifold, 91
Einstein metric, 91
energy, 69, 75, 81
energy functional, 26
equivariant map, 25
Euclidean Bruhat-Tits building,
     31, 45, 55, 81, 96
Euler-Lagrange equations, 26
existence of geodesics, 4
existence of minimizers, 76
exponential growth, 9
exponential map, 12, 14

Faltings, 22
Farrell, 9
first axiom of countability, 74
first Betti number, 5, 88, 92
first homology group, 5
flat, 17
Frankel, 82
fundamental group, 4

Γ-limit, 74, 75, 82
Gaussian kernel, 71
generalized harmonic map, 81
geodesic, 35, 43
geodesic arc, 3, 61
geodesic homotopy, 4, 52, 82
geodesic length space, 43, 61, 63, 72,
     73
geodesic quadrilateral, 43
geodesic space, 43
global, 46, 58
Grauert, 19, 21
Gromov, 6, 7, 10, 11, 18, 29, 42, 45,
     53, 81, 83, 97

h-component, 80
h-energy, 70

h-equivalent, 79
Hadamard-Cartan theorem, 4, 35
Hadamard manifold, 4
Harish-Chandra, 29
harmonic function, 72
harmonic map, 24, 27, 69, 80, 85, 86,
     88, 91, 92, 95–97
harmonic one-form, 88
Hartman, 83
Heintze, 7
Helgason, 17
Hermitian manifold, 3
higher homotopy groups, 4, 24
Hodge theorem, 88
holomorphic curve, 19
holomorphic map, 3
holomorphic sectional curvature, 2
homotopy equivalence, 24
Hopf-Rinow theorem, 35
Hummel, 7
Hunt, 22
Hurwitz, 5, 11
hyperbolic, 2, 3
hyperbolic Cayley plane, 5, 18, 29, 97
hyperbolic metric, 2
hyperbolic space, 5, 17
hyperbolicity, 2

Im Hof, 11
inner, 44
interior, 44
irreducible, 17, 23
isometry group, 6
isoperimetric inequality, 10
isotrivial, 19
isotropic, 3
isotropy group, 13

Jacobi field, 14, 33–35
Jones, 9
Jost, 18, 20–22, 33, 76, 80, 81, 87, 91,
     97

K3 surface, 22

Kähler-Einstein metric, 22
Kähler geometry, 2
Kaneyuki, 92, 94
Killing form, 12, 93
Killing vector field, 14
Kobayashi, 2, 21
Korevaar, 81, 97

Labourie, 76, 82
Laplace-Beltrami operator, 85
lattice, 28
Lebesgue space, 45
Lemaire, 85
length, 43
length space, 44
Levi-Civita connection, 14
Lie triple system, 17
local field, 28
locally symmetric spaces, 5, 6
Lohkamp's theorem, 2
Looijenga, 22
lower semicontinuous, 75

Makarov, 29
Manin, 19
mapping class group, 5, 6, 11
Margulis, 23, 27, 29, 83, 95, 96
Margulis superrigidity theory, 6
Matsushima, 92, 94
mean value, 65, 72
measurable, 69
Menger, 43
midpoint, 44
Milnor, 9
minimizer, 75, 82
minimizing point, 61
minimizing sequence, 62, 64, 78
moduli, 5
moduli space, 19, 21
Mok, 92
Mordell problem over function fields,
        19
Moreau-Yosida approximation,
        62, 76, 78

Mostow, 7, 23, 29
Mostow's rigidity theorem, 6, 9, 23
Mozes, 10
Mumford-Deligne
        compactification, 19

Nagano, 92, 94
negative (nonpositive) sectional
        curvature, 1
negative curvature, 3
negative curvature (in the sense of
        Busemann), 45
negative holomorphic sectional
        curvature, 3
negative isotropic sectional
        curvature, 3
negative operator, 3
negative Ricci curvature, 6
Nikolaev, 55
Noguchi, 21
nonarchimedean, 96
nonarithmetic lattice, 29
nonpositive sectional curvature,
        3, 34, 38, 41
nonuniform Busemann NPC space,
        50
nonuniform lattices, 18

Ohsawa, 81
Olshanski, 10
open ball, 43

parallel, 83
parallel family, 82
parametrized proportionally
        to arclength, 43
Parshin, 19
Peng, 20
Piatetski-Shapiro, 29
polynomial growth, 9
Prasad, 23
Preissmann's theorem, 4, 8
projection, 67
Pythagoras inequality, 58

quadrilateral, 49, 56
quaternionic hyperbolic space, 29

rank, 17, 28
rational, 28
Rauch comparison theorem,
        34, 38, 41
real hyperbolic space, 29
Reckziegel, 21
reductive, 78, 80, 81, 96
Reshetnyak, 56
residue field, 30
ρ-equivariant, 73, 76, 81
Ricci curvature, 2
Ricci tensor, 87, 91, 93, 94
Riemann moduli space, 5
Riemann surface, 5
Riemannian normal coordinates,
        86, 90
Rips, 10
Royden, 21

Sampson, 81, 89, 92
Schoen, 29, 81, 83, 97
Schroeder, 7, 18, 48
Schwarz lemma, 20, 21
Schwarz-Pick lemma, 2
second axiom of countability, 75
sectional curvature, 1
Sela, 10
semisimple, 12
semisimple Lie group, 11, 27, 28
Serre, 31
Shafarevitch problem over
        function fields, 19
shortest geodesic, 43, 73
sign conventions, 1
simple, 28, 69
simple ideal, 93
simply connected, 4
Siu, 7, 20, 92
splitting, 23
strictly convex, 36, 44, 46, 61, 81
structure constants, 93

superrigidity, 27
superrigidity theorem, 95
superrigidity theorem of Margulis, 28
Švarc, 9
symmetric space, 14, 16, 22, 78, 92,
        95, 97
symmetric spaces of noncompact
        type, 11, 17, 23

Teichmüller space, 5
theorem of Matsushima, 6
Thurston, 7
tight, 44
Todorov, 22
torsionfree, 18
torus, 4
totally disconnected, 28
totally geodesic, 17, 24, 89, 94–97
totally isotropic, 3
tree, 29, 31, 45
tripod, 65
Tromba, 20

uniform, 50
uniform lattice, 18, 23
uniqueness, 83, 98
uniqueness of geodesics, 4

valuation, 29
valuation ring, 30
Vesentini, 92
Vinberg, 29

Wald, 43
weighted mean value, 71
Weil-Petersson metric, 20
Wolpert, 20
word hyperbolic, 10

Yau, 21, 22, 76, 81, 91
Yeung, 92

Zariski closure, 28
Zariski dense, 28, 95
Zheng, 7
Zimmer, 27

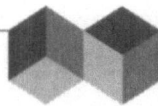

Mathematics with Birkhäuser

DIFFERENTIAL GEOMETRY • TOPOLOGY • GEOMETRIC GROUP THEORY

DMV Seminar 25

W. Ballmann, University of Bonn, Germany

Lectures on Spaces of Nonpositive Curvature

with an appendix by Misha Brin
Ergodicity of Geodesic Flows

1995. 112 pages. Softcover
ISBN 7643-5242-6

Singular spaces with upper curvature bounds and, in particular, spaces of nonpositive curvature, have been of interest in many fields, including geometric (and combinatorial) group theory, topology, dynamical systems and probability theory. In the first two chapters of the book, a concise introduction into these spaces is given, culminating in the Hadamard-Cartan theorem and the discussion of the ideal boundary at infinity for simply connected complete spaces of nonpositive curvature.

In the third chapter, qualitative properties of the geodesic flow on geodesically complete spaces of nonpositive curvature are discussed, as are random walks on groups of isometries of nonpositively curved spaces. The main class of spaces considered should be precisely complementary to symmetric spaces of higher rank and Euclidean buildings of dimension at least two (Rank Rigidity conjecture). In the smooth case, this is known and is the content of the Rank Rigidity theorem. An updated version of the proof of the latter theorem (in the smooth case) is presented in Chapter IV of the book. This chapter contains also a short introduction into the geometry of the unit tangent bundle of a Riemannian manifold and the basic facts about the geodesic flow.

In an appendix by Misha Brin, a self-contained and short proof of the ergodicity of the geodesic flow of a compact Riemannian manifold of negative curvature is given. The proof is elementary and should be accessible to the non-specialist. Some of the essential features and problems of the ergodic theory of smooth dynamical systems are discussed, and the appendix can serve as an introduction into this theory.

With a few exceptions, the book is self-contained and can be used as a text for a seminar or a reading course. Some acquaintance with basic notions and techniques from Riemannian geometry is helpful, in particular for Chapter IV.

"...This book is a fine introduction to the modern theory of spaces with curvature < 0 and is written by one of the leading researchers in this field....make it an attractive addition to one's library, whether one is an expert or simply a person who wants to know more about the recent developments in this area."

ZENTRALBLATT MATHEMATIK 1996

For orders originating from all over
the world except USA and Canada:
Birkhäuser Verlag AG
P.O Box 133
CH-4010 Basel/Switzerland
Fax: +41/61/205 07 92
e-mail: farnik@birkhauser.ch

For orders originating in the
USA and Canada:
Birkhäuser
333 Meadowland Parkway
USA-Secaurus, NJ 07094-2491
Fax: +1 201 348 4033
e-mail: orders@birkhauser.com

Birkhäuser

Birkhäuser Verlag AG
Basel · Boston · Berlin

VISIT OUR HOMEPAGE **http://www.birkhauser.ch**

Mathematics with Birkhäuser

BAT • Birkhäuser Advanced Texts / Basler Lehrbücher

L. Conlon, Washington University, Saint Louis, MO, USA

Differentiable Manifolds
A First Course

1992. 369 pages. Hardcover.
First edition, 2nd revised printing
ISBN 3-7643-3626-9

The basics of differentiable manifolds, global calculus, differential geometry, and related topics constitute a core of information essential for the first or second year graduate student preparing for advanced courses in differential topology and geometry. Differentiable Manifolds is a text designed to cover this material in a careful and sufficiently detailed manner, presupposing only a good foundation in general topology, calculus, and modern algebra. It is ideal for a full year PhD qualifying course and sufficiently self contained for private study by non-specialists wishing to survey the topic.

The themes of linearization, (re) integration, and global versus local calculus are emphasized repeatedly; additional features include a treatment of the elements of multivariable calculus, an exploration of bundle theory, and a further development of Lie theory than is customary in textbooks at this level.

Students, teachers, and professionals in mathematics and mathematical physics should find this a most stimulating and useful text.

CONTENTS: TOPOLOGICAL MANIFOLDS • THE LOCAL THEORY OF SMOOTH FUNCTIONS • THE GLOBAL THEORY OF SMOOTH FUNCTIONS •FLOWS AND FOLIATIONS • LIE GROUPS AND LIE ALGEBRAS • CONVECTORS AND 1-FORMS • MULTILINEAR ALGEBRA AND TENSORS • INTEGRATION OF FORMS AND DE RHAM COHOMOLOGY •FORMS AND FOLIATIONS • RIEMANNIAN GEOMETRY • INDEX

"... Conlon's book serves very well as a professional reference, providing an appropriate level of detail throughout. Recommended for advanced graduate students and above."

S.J. Colley, Oberlin College

For orders originating from all over the world except USA and Canada:
Birkhäuser Verlag AG
P.O Box 133
CH-4010 Basel/Switzerland
Fax: +41/61/205 07 92
e-mail: farnik@birkhauser.ch

For orders originating in the USA and Canada:
Birkhäuser
333 Meadowland Parkway
USA-Secaurus, NJ 07094-2491
Fax: +1 201 348 4033
e-mail: orders@birkhauser.com

Birkhäuser
Birkhäuser Verlag AG
Basel · Boston · Berlin

VISIT OUR HOMEPAGE **http://www.birkhauser.ch**

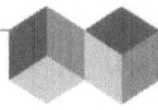
DIFFERENTIAL GEOMETRY • TOPOLOGY

MMA 90
Monographs in Mathematics

Ph. Tondeur, University of Illinois, Urbana, IL, USA

Geometry of Foliations

1997. Approx. 312 pages. Hardcover
ISBN 3-7643-5741-X
due Summer 1997

This volume describes research on the differential geometry of foliations, in particular Riemannian foliations, done over the last few
years. It can be read by graduate students and researchers with a
background in differential geometry and Riemannian geometry. Of
particular interest will be the Hodge theory for the transversal Lapla-
cian, and applications of the heat equation method to Riemannian
foliations.

There are chapters on the spectral theory for Riemannian foliations,
on Connesí point of view of foliations as examples of noncommutative
spaces, and a chapter on infinite-dimensional examples of Riemannian foliations.

For orders originating from all over
the world except USA and Canada:
Birkhäuser Verlag AG
P.O Box 133
CH-4010 Basel/Switzerland
Fax: +41/61/205 07 92
e-mail: farník@birkhauser.ch

For orders originating in the
USA and Canada:
Birkhäuser
333 Meadowland Parkway
USA-Secaurus, NJ 07094-2491
Fax: +1 201 348 4033
e-mail: orders@birkhauser.com

Birkhäuser

Birkhäuser Verlag AG
Basel · Boston · Berlin

VISIT OUR HOMEPAGE **http://www.birkhauser.ch**